THE DIARIES OF ROBERT HOOKE,
THE LEONARDO OF LONDON,
1635–1703

THE DIARIES OF ROBERT HOOKE,
THE LEONARDO OF LONDON, 1635–1703

Richard Nichols

The Book Guild Ltd.
Sussex, England

This book is sold subject to the condition that it shall not, by way of trade or otherwise, be lent, re-sold, hired out, photocopied or held in any retrieval system or otherwise circulated without the publisher's prior consent in any form of binding or cover other than that in which this is published and without a similar condition including this condition being imposed on the subsequent purchaser.

The Book Guild Ltd.
25 High Street,
Lewes, Sussex

First published 1994
© Richard Nichols 1994
Set in Times
Typesetting by Acorn Bookwork, Salisbury, Wiltshire

Printed in Great Britain by
Antony Rowe Ltd,
Chippenham, Wiltshire.

A catalogue record for this book is
available from the British Library

ISBN 0 86332 930 6

CONTENTS

INTRODUCTION
The Renaissance of Science in Britain and the founding of the Royal Society   1

HOOKE THE MAN
Early Life   13
Domestic Life   23
Dish of the Day   29
Brother John Hooke   35
A Summary of Some Details of Hooke's Life from the Diary   38

ROBERT HOOKE – THE LEONARDO OF LONDON
Hooke's *Micrographia*   51
Meteorology   56
Comets   60
Astronomy   62
Investigations into Air   64
Clockmaking and the Longitude   78
Experimenting in Brick Making   85
Measuring a Degree and the Circumference of the Earth   87
Geology   93
The Waywiser   98

Surveyor and Architect	101
A new Building for The Royal Society	110
Blood Transfusion and Skin Grafting	114
A Universal Language	120

HOOKE'S COLLEAGUES AND CONTEMPORARIES

John Aubrey	125
Sir William Petty	130
Denis Papin	133
Isaac Newton	135
Sir Christopher Wren	142
Charles II	147
Sir John Cutler	150

CONCLUSION

Nullus in Verba	157
Postscript	160

APPENDIX

Music and Sound	163
Shipping	164
A Perfect Wheel Work	166
New Kinds of Levels	166
An Alphabet of Symbols	167
The Universal Joint	167
A Device for Speedy Intelligence	168
Education	168

Lamps	169
Flying	169
Some Random Experiments Not Pursued	172
A Few Suggested Cures of the Time	173
Buildings designed by Hooke	181
BIBLIOGRAPHY	184

PREFACE

Sir Frederick Gowland Hopkins, O.M., President of the Royal Society, wrote the foreword to the publication of Hooke's *Diary* in 1935. He stated that it was a vivid record of the scientific, artistic and social activities of a remarkable man.

Hooke's scientific activities were associated with his position as the first Curator of the Royal Society, an office he held from 1662 until his death in 1703. A study of the experimental work done during this period shows it has its peak during the years 1662–1665, and much of this was summarised in Hooke's first publication, *Micrographia*. The *Diary* recorded in brief, laconic statements his other scientific work; his theory of springs, the principle of the arch, ideas for improving telescopes, barometers, clocks and watches, carriages and shipping, and the craftsmen he employed to make his scientific instruments.

The rebuilding of the City of London after the Great Fire gave Hooke the opportunity to show his skill as a surveyor and architect, his ability as a draughtsman, and his scientific knowledge about materials. His position as Curator of The Royal Society led to his introduction to all ranks of society, from the King and his ministers, to the Lord Mayor of London, from the nobility to fellow craftsmen and instrument makers. There are more than two thousand people listed in the index of the *Diary*, ranging from the King to a one-eyed sailor. Hooke was a very gregarious man, and interested in hearing and discussing the experiences of all the people he met in the coffee houses and inns of the City.

Details of his diet and medication are given in the *Diary*. He could be considered as a hypochondriac from his interest in urinology, his stools and the condition of his blood, but these were the standard indicators of health or disease in his day. The liquids which he drank, the spa waters sold in the Strand are all noted, as is their effectiveness in promoting his health.

The fees he received from surveying are noted, as well as those from his building commissions. These gave him the means to add to the books bequeathed to him by his father, and build up a library of more than three thousand books. He was also able to buy cloth from Ravenscroft and Blennerhasset for his housekeeper, Nell, to make up into suits, and to buy a sword and pistols.

There is a great deal of scientific detail in the diary, as well as a wealth of social history. The object of this book has been to bring these details together, presenting them in a readable way, and linking them to his other works. This should lead to a proper appreciation of his contribution to scientific progress, and to the inventions from which we benefit today.

R.N.

INTRODUCTION

THE RENAISSANCE OF SCIENCE IN BRITAIN AND THE FOUNDING OF THE ROYAL SOCIETY

The seventeenth century saw the advent of a new approach to science which was to bring great advances in all fields of inquiry. The main inspiration for this was Francis Bacon, 1st Lord Verulum. The methods that he advocated had been used by William Gilbert in his experiments in magnetism, and by William Harvey in his research by which he demonstrated that the blood must circulate around the body.

Harvey's opinion of Bacon's writings was that he wrote philosophy like a Lord Chancellor, but Bacon's idea was that the work and methods of inquiry used by Aristotle should be abandoned. He advocated the collection of facts and the selection of those which seemed to be relevant, devising a hypothesis and testing this by experiments from which a theory could be established. This was to be the method used by members of the Royal Society, and expressed in its motto, *Nullus in Verba*, Not by Word Alone.

Francis Bacon died in the cause of science by testing whether a chicken could be preserved by stuffing it with snow. Unfortunately he caught a chill, and died at the house of his friend, the 2nd Earl of Arundel, at the Old Hall, Highgate.

One of the founder Fellows of the Royal Society, whose life extended over this period of change was John Evelyn. He was born on 31st October 1620, and died on 27th February 1706. He had spent several years on the Continent during the time of the Civil Wars and visited Italy, France and the Netherlands, returning to England on

6th February 1652. He was to attempt to provide some of the ideas which had developed during these journeys, and saw some of them brought to fruition, such as the use of bricks for house building, as he had seen in the Netherlands. He suggested ways of reducing the nuisance of smoke in London, schemes for re-afforestation and the composting of waste material. His diary records his interest in the founding of the Royal Society, and his participation in its organisation. He was probably responsible for the motto which it adopted and its title. He was Secretary of the Society for a year, and a member of its council on several occasions over a period of thirty years.

One of the first mentions of Hooke in Evelyn's diary was on 4th March 1664 – "Came to dine with me the Earl of Lauderdale, his Majesty's great favourite, and Secretary for Scotland; The Earl of Teviot; my Lord Viscount Brouncker, President of the Royal Society; Dr Wilkins, Dean of Ripon; Sir Robert Murray, and Mr Hooke, Curator of the Society."

The other diarist who was a contemporary of Hooke was Samuel Pepys. He was elected a Fellow of the Royal Society on 15th February, 1665. In January Pepys had bought a copy of Hooke's *Micrographia*, and sat up till 2 in the morning reading it. He also recorded details from some of Hooke's lectures, demonstrations and conversations. Pepys had not had the breadth of education of Hooke and, although he accepted Hooke's explanations, such as the method of determining the number of beats of a fly's wing, Pepys had to have lessons in arithmetic, even to learning his multiplication tables.

He acquainted himself very thoroughly with all matters relating to the building and provisioning of ships, and was the acknowledged expert in these matters, as he demonstrated in his speech to Parliament in his defence of his administration. He also was a member of the council of the Royal Society on several occasions, and President in 1684–6. Of the first two hundred Fellows of the Royal Society, these two diarists were active members, whereas 126 of the

Fellows were either inactive or only slightly active, according to the study made by Dr Michael Hunter of the Society.

Robert Hooke was Curator of the Royal Society from 1663 until his death in 1703, and during this period made the most numerous contributions to its meetings.

In the opinion of Marjorie Hope-Nicholson, an American historian of science, expressed in her book, *Pepys' Diary and the New Science*, Hooke *was* the Royal Society.

The object of this work is to remove from obscurity the details of the life of this remarkable man. He was widely read, had been educated by the foremost scholars of his day and, during the period when the diaries were written, was in regular contact with them and discussed at any opportunity the wide range of subjects raised at meetings at The Royal Society, and of topical concern. His inspiration was the philosophy of Francis Bacon, whom he referred to as The Noble Verulam, and who was also the inspiration for the founders of The Royal Society. His early patrons were, "the great Ornament of our Church and nation", the Lord Bishop of Exeter, (Dr Seth Ward), Dr Wilkins, "a man born for the good of mankind", and for the honour of his country, and to whom he attributed the encouragement and promotions of Hooke's own interests, and the most illustrious Mr Boyle, his particular patron, and the patron of philosophy itself.

Hooke was fulsome in his praise for these people, but that age had many notable scientists who made important discoveries, and raised the esteem of Britain in the eyes of the scientific community of Europe.

Hooke's several contributions to the development of scientific progress are described in this book with details of social interest about the era in which he lived.

There are at least 200 papers, books or articles in scientific journals relating to Hooke, his work and influence and associations with other scientists of his day. He was conscious of his own shortcomings on occasion, as when in

1676 he wrote, "Resolve more diligence and resolution". Another comment made which he noted in his diary was that he was minding business and profit too much, and Sir Joseph Williamson remarked that he wanted a higher chair. As Hooke had been in association with the founding fathers of the Royal Society, first as a scholar at Oxford, then assistant to Dr Willis and Robert Boyle, and on the founding of the society, was constantly referred to for his opinion and for his help in pursuing their interests, he no doubt had a high opinion of himself, but as these men were the foremost scientists of their day, and respected his ability, and achievements, and the part he played in establishing the respect gained by other learned bodies for the Society, as Curator of Experiments from 1663 to 1703, he had good reason to have some self esteem. This summary of a few of his works justifies that. Many of them are still in use today, and should be recognised.

THE FOUNDING OF THE ROYAL SOCIETY

Birch related that most of the group of scientists who had met at Oxford returned to London about the year 1659. They held their meetings at Gresham College, after Wren's lecture on astronomy on Wednesdays, and on Thursdays after Rooke's lecture on geometry. After the Restoration of Charles II, they continued their meetings and, on that held on 28th November 1660, they considered founding a college for promoting physico-mathematical learning. A week later the group were informed that the King approved of their design and would be ready to give encouragement to it.

On 10th April 1661, Hooke's tract on water rising to different heights in glass tubes of different bore was ordered to be read and this was done, the water rising to the greatest height in the narrowest glass tube.

On 28th May 1662, the Bishop of Exeter related the intention of a person of quality to assist the Society in their experiments. The Charter incorporating the Society under the title "The Royal Society" was passed. One of its provisions was the appointment of a Curator of Experiments, and on 5th November of that year, Sir Robert Moray proposed that Robert Hooke, a person offering to furnish them every day on which they met with three or four considerable experiments and expecting no recompense till the society should get a stock enabling them to give it, be employed. The proposition was received unanimously, and it was ordered that Mr Hooke should come and sit amongst them, and both bring in every day of their meeting three or four experiments of his own, and take care of such others as should be mentioned to him by the society.

Among the experiments conducted by Hooke after this

date were weighing common and rarefied air; estimating the height of the atmosphere; weighing bodies at the top of Westminster Abbey; studying the effect of exhausting the air from a receiver in which was a tench (its eyes bulged out) and the speed with which bodies fell in a vacuum, to name but a few.

During the same period Sir Robert Moray suggested studying the effect of pre-heated air being blown into a furnace, and he had the idea of using black lead, graphite, as a lubricant for engines, an idea suggested by Christopher Wren previously as the lubricant for watches. Wren had also suggested the oscillation of a pendulum as a standard of length. Robert Boyle had read a paper on potatoes, and given some to members to plant and report their results. Hooke had been asked to examine the water from Barnet well and reported that there were creatures swimming in it. He showed several of his other microscopia. Dr Goddard gave Hooke a piece of petrified wood, which Hooke cut and polished, and found its specific gravity as 3.5. It was as hard as flint, would readily cut glass, would strike a spark against steel, would not burn, and yielded bubbles when vinegar was put on it.

Hooke was elected a Fellow of the Royal Society on 3rd June 1663, and exempted from all charges.

He showed the artificial eye that he had made on 12th October 1663 which included the iris diaphragm made of overlapping plates by which the area of the iris could be varied. He was asked to develop an engine to kill whales. These are a few of the several tasks he was asked to carry out for the Fellows.

He also showed some of his examinations of objects viewed through his microscope, and he was directed to prepare a book to publish them. This he did, and the book, *Micrographia*, was ready for sale in January 1665.

Hooke had been elected Professor of Geometry at Gresham College on 8th June 1664, on the casting vote of the Lord Mayor, and on 15th June he had been proposed as Cutlerian Lecturer in the Mechanical Arts at a salary of

£50 per annum. The Royal Society decreased his previously agreed salary of £80 per annum by that amount.

In 1665 when the death rate from the plague increased greatly, the Royal Society stopped its meetings in London and Fellows dispersed to safer places in the country. Hooke, Dr Wilkins and Sir William Petty went to Durdans, near Epsom.

The meetings resumed in February 1666.

HOOKE THE MAN

EARLY LIFE

Robert Hooke was born on 18th July 1635 at 12 noon, at Freshwater, Isle of Wight. His father, John Hooke, had been curate of the church of All Saints, Freshwater, since at least 1626. Waller used the term minister to describe him. The Rector at that time was the Reverend George Warburton, described as of Cheshire, Gent., who had been Chaplain-in-Ordinary to James I, and to Charles I, as well as being Dean of Gloucester Cathedral and of Wells. John Hooke was left to minister to the spiritual needs of the parishioners, as well as teaching at the school associated with the church. He was also tutor to the children of Sir John Oglander, one of the principal landowners in the island. All Saints was a wealthy living in the patronage of St John's College, Cambridge. The wealth of this living is noted, as by some means John Hooke was able to leave £100 to his son Robert at his death, and the older son John was able to set up as a grocer at Newport. The stipend of lesser clergymen was £40–£50 per annum.[1]

The few details of Robert's early life are recorded in Aubrey's *Brief Lives*, and by Richard Waller in his introduction to the *Posthumous Works of Robert Hooke*, first published in 1705. These details would have been given by Hooke to his two friends during their many meetings in coffee houses and in Hooke's rooms at Gresham College.

From these accounts, Robert was a sickly child from birth, and for the first seven years his parents doubted

---

[1] The correct amount of money bequeathed to Robert Hooke by his father was £40, together with all his father's books.

The Rev John Hooke's will was discovered in the Hampshire Record Office by Hideto Nakajima of Tokyo University, and first published in The Robert Hooke Society Newsletter of April, 1993.

At his death Hooke's library totalled 3,200 books, which were auctioned as was customary at that time. Sir Jonas Moore's numbered, 2,000 and Newton's 1,900.

whether he would survive. As Maurice Ashley records in his *Life in Stuart England*, "Many children were born, but few survived". In spite of this he was described as being very sprightly and active in running and leaping, though very weak as to any robust exercise. His father had intended that Robert should be trained for the Church, and took some pains over his education, but Robert's apparent weak constitution, and his father's increasing ill health, ended this ambition. Robert was left to his own devices, and was not set on the path of classical studies, a method criticised by Milton in his *Essay on Education*, in which he stated, "First we do amiss to spend seven or eight years in scraping together so much miserable Latin and Greek as might be learned easily and delightfully in one year".

In a small town by the sea there was much to arouse the interest of a lively boy. The shells and rocks on the sea shore, the cliffs with their varied formations, and a cave well above high-water mark to explore. Fishing vessels, merchant ships and warships making their way through the Solent, and at times striving to make headway against the strong tidal currents, were all subjects for study. The surrounding countryside with its farms, crops and animals, the wild flowers and trees were all full of interest for an observant boy.

Robert soon showed his inventive ability and mechanical skill. He made a model warship a yard long, complete with rigging, sails and guns which could be fired. Noticing an old brass clock which had been taken to pieces, he copied the parts in wood, and constructed a clock which worked, a feat which John Harrison, the maker of the ship's chronometer, was also to do later. Hooke made a dial on a trencher, a simple form of sun dial. "John Hoskyns, the portrait painter, being then at Freshwater, Hooke watched him at work, and then got chalk and ruddle as pigments, put them in a trencher, got a pencil, and to work he went, and made a picture; then he copied (as they hung up in the parlour) the pictures there, which he made like." Hooke himself recorded that he copied several prints with a pen, "which Mr. Hoskyns (son of the famous Hoskyns, Cowpers master) so much admired as one not instructed could so well imitate them."

Hooke's father died in 1648, having been in bad health for three or four years and, as Robert had shown some ability at drawing, he was sent with the £100 legacy to London to be apprenticed to Peter Lely, the portrait painter.

In the words of John Aubrey: "but Mr Hooke perceived what was to be done, so thought he, why cannot I do this myselfe and keep my hundred pounds?".

So he left Peter Lely's studio, and enrolled as a pupil at Westminster School, where he lodged in Dr Busby's house.

In his book *The King's Nurseries; The Story of Westminster School*, John Field, the archivist of the school, refers to Dr Busby as justly the most celebrated head master in Westminster's long history, and possibly in the whole history of education. It was said of him that he knew all there was to know. He was also a formidable person in other respects. John Field relates how in 1641 he bought twenty muskets and twenty pounds each of powder and shot, and in the following year the Westminster boys helped to defend the Abbey against a mob intent on pulling down the organ and monuments.

On the day King Charles was executed, Dr Busby led the school in prayers for the King. In 1644 when an Ordinance was passed for the demolition of monuments of idolatory and superstition and the removing of organs from churches, Dr Busby must have saved that of the Abbey, and he also had one in his own house. An old scholar, Edward Wetenhall recorded that the first organ that he ever saw or heard was in Dr Busby's house where it was used regularly for services.

While he was a scholar at Westminster, Hooke learned to play twenty four lessons on the organ. He also gained Dr Busby's respect by mastering the first six books of Euclid in his first week at school. Dr Busby seems to have recognised that he had an exceptional pupil. Sir Richard Knight, a contemporary of Hooke, later stated that he was rarely seen in school, and was probably left to study by himself in Dr Busby's library. John Locke, who was also a contemporary of Hooke, later related in one of his essays on education "that he knew a young gentleman, bred something after this way [in geometry] able to demonstrate several propositions in Euclid, before he was thirteen". He does not give his name but he was almost certainly refering to Hooke. Locke was also one of Dr Busby's favourite pupils, his "White Boys".

Dr Busby was a noted disciplinarian, but one incident showed that he appreciated a sharp wit. A boy was called out for some misdemeanour and Dr Busby said he would marry him to his rod. The boy protested that the banns had not been called three times, so the Doctor excused him from being birched.

Locke's opinion of beating was that it should only be inflicted for rebellion or obstinacy. Like Milton, he also thought that labour for labour's sake is against nature, and that curiosity should be carefully cherished in children, as other appetites are suppressed.

Hooke was fortunate in gaining the respect of Dr Busby and being left to follow his own pursuits of knowledge just as he had before attending Westminster School. [Aubrey stated that Hooke invented "thirty several" ways of flying, but these claims would have been from Hooke's own memoirs given to Aubrey.] He became a proficient Latin and Greek scholar as is evident from the depth of knowledge shown in the lectures published in his *Posthumous Works*. He also enjoyed having conversations in Latin with Theodore Haak, on the many occasions when that friend and chess player came to visit him later at Gresham College.

He expressed himself naturally in Latin when feeling pleased, as on 1st July 1680, when after sleeping well, he wrote *Soli Deo Gloria*, and on attaining his 45th birthday on 18th July that year, *Deo Gratias*; while, when he was annoyed, as after seeing the play, *The Virtuoso*, he exclaimed "Damned Doggs. *Vindica me Deus.*"

Hooke's association with Dr Busby continued after Hooke had left Westminster for Oxford University and, after 1672, when Dr Busby was made Archdeacon of Westminster, Hooke carried out several commissions for him. He re-paved the choir of the Abbey, at a cost of £3, and designed Dr Busby's library. This was damaged in an air raid during 1939–45, but has been restored according to original drawings made of it. The black and white marble pavement of the choir escaped damage, and can be seen on state occasions within the Abbey.

Dr Busby was "sickly" for having to take the oath of the Covenant, and walked in Cromwell's funeral procession as one of the dignitaries associated with Westminster, but one of his pupils, Robert Uvedale, upheld the traditions of the School in its loyalty to the monarchy, by pushing through the ranks of soldiers lining the route and seizing the Majesty Escutcheon placed on the bier and escaping through the crowd. This is now a treasured possession of the School.

Hooke's Studies at Oxford

In 1653 Hooke was entered as a chorister at Christ Church, Oxford, which Aubrey stated, was "a pretty good maintenance". He was in the chamber of Mr Burton, author of *The Anatomy of Melancholy*, of whom it was stated that "non obstante, all his Astrologie and book of Melancholy, he ended his days by hanging himself." Burton had predicted the date of his death from that of his birth, but there is no mention of this as to the cause of his death in the preface of his book in the Everyman edition. He died on 25th January 1640 and was buried in the Cathedral at Oxford, which would be unlikely if he had committed suicide. He left his library of 2,000 books to the College and the Bodleian. The Bodleian was saved from damage when Oxford was captured by the parliamentarian army, by the prompt action of Thomas Fairfax, its General.

With the re-organisation of the university after the civil wars, a number of eminent scholars were installed in the university. Dr Wilkins, Cromwell's brother-in-law, was made Warden of Wadham College. Dr Wallis was made Savillian Professor of Geometry, and Dr Seth Ward Savillian Professor of Astronomy. Dr Petty and Dr Willis were also appointed to university posts.

As Godfrey Davies wrote in *The Early Stuarts*, "In spite of wholesale expulsions and the distractions of the times, the standard of scholarship at the universities was well maintained and discipline much improved."

Even Clarendon, a loyal son of Oxford, voiced a tribute to, "the harvest of extraordinary good and sound scholarship in all parts of learning: the men who were introduced applied themselves to the study of good learning and the practice of piety."

In 1649 Cromwell had said, "no Commonwealth could flourish without learning".

He presented the Bodleian Library with some Greek manuscripts and, on his election as Chancellor, had appointed Dr Owen, his favourite chaplain, as Vice-Chancellor.

According to Dr John Wallis' account of the origins of the Royal Society given in Birch's *History of the Royal Society*, Dr Wallis, Dr Wilkins, Dr Goddard and Dr Petty were among the group of several worthy persons who had met at Dr Goddard's lodging in Wood Street, Cheapside, London in about the year 1645 to discuss subjects associated with the new experimental philosophy, but avoiding any political or religious matters. After their appointment to academic posts at Oxford, these men continued their meeting, first at Dr Petty's lodgings and, after his posting to Ireland, at Dr Wilkins' at Wadham College.

Cromwell's interest in higher education continued throughout his life, and one of his last acts was the founding of Durham University. When Oliver died, the University of Cambridge petitioned Richard Cromwell to annul this foundation, which he did. In a footnote with respect to this act, in *The Diary of Thomas Burton*, it is mentioned that Dr Petty favoured the founding of a university of London, to give anyone the opportunity of advancing himself in life, as he had done.

Hooke had the benefit of being instructed by and joining in the discussions of these men. He learnt dissection from Dr Willis, and astronomy from Dr Seth Ward. Dr Wilkins gave him a copy of his book, *Mathematical Magick*, the first part of which is on mechanics and deals with the principles of levers, pulleys and what can be achieved using machines based on these. It also includes Dr Wilkins' ideas about the possibility of flying and travelling to the moon.

Hooke continued his experiments on flying in the grounds of Wadham College, but after trying various ideas based on pulleys he concluded flying was not possible by such means. The subject was to occupy his thoughts and discussions with Wren for many of the years covered by the diary. He considered making artificial muscles, having decided that the muscles of a man's body were not sufficient for the purpose. Waller described this group of scholars as, "that concourse of extraordinary persons at Oxford, each of which afterwards were particularly distinguished for the great light they gave to the learned world by their labours."

Dr Willis recommended Hooke to Robert Boyle as his assistant, and so began an association and lifelong friendship. Hooke's first and undisputed invention was that of the improved air pump, by which means many of the physical and chemical properties of air were investigated. Boyle's interest in this device had been aroused by reading of Von Guericke's experiments. A better pump had been made by Greatorix, but Hooke thought this, "too gross to perform any great matter". The one invented by Hooke could evacuate the air much faster from the receiver, the receiver could be put on a separate platform, and receivers could be made large enough for a man to sit inside and experience the effects of reduced air pressure, an experience to which Hooke later subjected himself.

Hooke's first notable discovery with Boyle was the establishment of Boyle's Law, for which, as Boyle was short-sighted and not good at calculations, Hooke probably took the readings and made the calculations.

During his time at Oxford Hooke also continued his interest in clocks. His studies in astronomy with Dr Seth Ward would have made the importance of accurate time-keeping an essential subject for that science.

The first invention to improve clocks was that of the anchor escapement. Dr Gunter, quoting J. E. Haswell in his book on horology, gives the date of this as 1656. This superseded the verge or crown wheel. Hooke's other invention was the application of a spring to the arbor of the balance wheel using the natural oscillation of the spring as a regulator, so giving greater accuracy to the clock, and also enabling it to be used in any position. This was of particular importance for the attempts to determine longitude at sea: another problem which was to occupy Hooke's attention, and which he was asked by the King to make efforts to solve. Priority to this invention was later disputed by Huyghens, but Dr Derham, who knew Hooke well, supported Hooke's claim, as did Mr Henry Sully, an English clockmaker living in Paris.

Huyghens had shown that Galileo's idea of adapting a pendulum to regulate a clock could be done, and this was quickly put into practice by clockmakers. Ahasuerus Fromantel was the first man to make pendulum clocks in England, and the one owned by Oliver Cromwell cost three hundred pounds, and only required winding once a month. Charles may well have retained this at his restoration, and it could have been the one against which Charles checked the accuracy of the watch invented by Hooke and made by Thomas Tompion.

Hooke had presented the King with the watch on 17th May, 1675, and on the following day he met the King in the park, who affirmed that the watch was very good. Hooke had demonstrated the spring-regulated watch to Lord Brouncker and others, and was advised to patent the invention, but because a clause in the proposed patent would have given anyone making an improvement on this to have the right of royalties, Hooke declined to patent it.

During the period covered by the first diary, from 1672–1680, Hooke was in very frequent association with Thomas Tompion, giving him the benefit of his many ideas for the improvement of the accuracy of watches, and of other instruments such as barometers and quadrants.

From an account in the *Philosophical Experiments and Observation of Robert Hooke* compiled by William Derham, some trials were made in 1662 of using pendulum clocks at sea to test their suitability for determining longitude. This was done by Lord Kincardine who had two pendulum clocks fitted below decks at right angles to each other. The cases were made heavier with lead and suspended by ball-and-socket joints. The clocks worked without stopping during the voyage, but did vary with each other.

Sir Robert Holmes also took two similar clocks on sea trials, and found that he was able to judge his ship's position more accurately than masters using traditional methods. Hooke's experience on the first voyage may have inspired him to invent a more reliable clock; a paper written by him in 1663 describing a marine clock was discovered in the library of Trinity College, Cambridge by Professor Rupert Hall three hundred years later. A model of this was made by Michael Wright of the Science Museum, and displayed in working order at a conference on Robert Hooke held at the Royal Society in July 1987.

Hooke was to continue his interest in all aspects of shipping, such as determining longitude at sea, false keels to give greater stability to ships, life belts, lighthouses, and way wisers for use at sea to determine the distance travelled.

When during the Great Plague the Royal Society suspended its meetings at Gresham College and most of the Fellows went into the country to avoid contagion, Hooke, Dr Wilkins and Dr Petty stayed at Durdans, near Epsom, and here they were visited by John Evelyn on 4th August 1665. The entry in Evelyn's Diary reads, "On my return, I called at Durdans, where I found Dr Wilkins, Sir William Petty and Mr Hooke, contriving chariots, new rigging for ships, a wheel for one to run races in, and other mechanical inventions; perhaps three such persons together were not to be found elsewhere in Europe for parts and ingenuity."

Evelyn reported that deaths from plague during the next two weeks in London as 4,000 and 5,000.

DOMESTIC LIFE

Hooke's appointment as Gresham Professor of Geometry gave him an income of £50 a year, rooms at Gresham College, and the task of giving a lecture a week during term time, once in Latin and again in English. The professors had also to be unmarried, and in the event of them marrying they had to relinquish their professorship. There was no bar against having a housekeeper, and the first of whom we hear was Nell Young, and the source of the detail is Hooke's diaries, the first kept between the years 1672 to 1680, and the second intermittently from 1688 to August 1693. Neither diary has had a wide circulation, the first and longer one because it consists mainly of unconnected jottings of the many and varied tasks he performed during those years, and the second because it is the last volume of the ten volume work of *Early Science in Oxford*, by Dr R.T. Gunter. The problem of deciphering the symbols used in these diaries must have presented a difficulty to the editors, but in the first diary the symbol ᛉ was taken to mean an orgasm, and Hooke satisfied his natural impulses with the co-operation of several of his housekeepers, Nell Young being the first. A few entries from the diary show Nell's varied tasks.

12th September 1672	"Nell lay."
29th September 1672	"Paid Nell for last quarter 20s."
8th October 1672	"ᛉ Nell at door."
28th October 1672	"Played with Nell hurt small of back."
2nd November 1672	"Nell made stript wastcot."
15th November 1672	"Gave Nell 30s for Graces clothes."
2nd December 1672	"Slept ill with Nell supra ᛉ."

These few entries show that Nell prepared his meals, made his clothes and those of Grace, and satisfied his other needs for £4 per annum. There are 205 entries relating to Nell in the first diary noting these various duties. Nell was married on 31st August 1673, and finally left his household on 31st August when she left to live with her husband at their home near to the Fleet Ditch which was being developed. She continued to visit Hooke, buying materials for his clothes and making them up. He often cut out the breeches and coats for her to sew. He also visited her, sometimes to have a shave, haircut and a cup of cocoa, for which she charged 6d. In the second shorter diary there are only 24 entries, and these relate the visits of Nell and her two daughters to see Hooke, and how she continued to make his clothes. There are no $\mathfrak{X}$ signs in this diary, perhaps because a different editor may not have considered there was any significance in the sign, or because Grace Hooke, who had been his housekeeper from 5th September 1675, until her death in 1687, and to whom he was greatly attached, was the last to appeal to him in this way.

Nell was succeeded as housekeeper by her cousin Bridget Taylor, who took over her duties in October 1673, but left at the end of the month. Hooke had also hired Dol Lord on 16th October 1673, who was not very satisfactory. He reproved her for her dressing, and once suspected that something had been lost when he found her sweeping his room when his cupboard door was open. In April 1674 he employed Bette Orchard at £3 per annum. He reproved her for burning a candle at night, and also when she broke one of his glasses. He "wrastled" with Bette on 20th June, and on 6th July wrote in his diary that, "Lost my cock by Bette's carelessnesse." He complained that she was intolerably lazy, and finally he paid her 15s on 30th September, and discharged her. He then employed Mary at 20s a quarter, and there was only one complaint against her when she broke a glass.

Hooke's niece Grace was first mentioned on 25th August 1672, when she dined with Hooke. On the following day he took her to Bartholomew Fair, where he spent 3s 6d on the sideshows. At a visit to the same fair in 1677 Hooke records seeing the tiger for which he paid 2d, the child doing strange tricks, and the Dutch woman, which exhibits were described in a poem entitled, *Wit and Drollery*, in which dancing bears were also mentioned. Four days later he bought a red pendant from Pargeter.

On 2nd September he sent her to school, and later he had some trouble with Bloodworth, "when he resolved to have Grace".

Grace returned to her father by the Portsmouth Coach on 25th November 1672, and although Hooke bought a pair of red pendants for Grace at a cost of 6s and sent them to her by coach on 4th January 1674, she did not return to London until September 1675, when she remained with Hooke until her death in 1687, except for an interval of a few months in 1677. In May 1676 he took her to the column to see the great fire of Southwark which destroyed eight or nine hundred houses. He also used to take her for walks to Islington, and paid for her and Tom Gyles to go to see *Antony and Cleopatra*.

Soon after her return to London in 1675 he took her to see the Lord Mayor's Show on 29th October, in the company of Mr Fitch, when they saw the unusual spectacle of the Lord Mayor falling off the back of his horse in the presence of the King. Hooke continued his interest in Grace's education, and wrote to his brother to ask if she could learn French, and he lent her a French tutor as a start. He also gave her and Tom Gyles lessons in algebra, and in the globes and maps. He also taught her bookbinding.

It was not until a year after her arrival that Hooke first slept with her, and another few months before the symbol Ӿ appears in his diary. His satisfaction with the arrangement is summed up in the entry for 5th March 1677, "Grace perfecte intime omne Ӿ. Slept well."

Grace returned to her family in August 1677, and six months later Hooke heard from Mr Young of Plymouth that Grace had measles. Hooke's brother John died in February 1678, and in June of that year Grace returned to Gresham College.

A year later Grace was taken ill with smallpox, and Hooke was most concerned. "Grace is desparately sick.", and, on the following day, "Grace exceedingly dangerously ill." She was visited by a Mr Whitchurch, who sent a dose of Gascoyne powder which made her sweat, and brought out the smallpox thicker. Mr Whitchurch also sent her the diascord cordial. Hooke wrote to Grace's mother who arrived five days later. He also got Dr Whistler, a Fellow of the Royal Society, to see Grace and he prescribed a cordial. Fortunately Grace recovered completely, and Hooke paid Dr Whistler his bill of 7s and gave him another 5s to buy a pair of gloves.

After her recovery she bound a book, and a week later completed the work of binding the two volumes of *China and Zimmerman*. Whitchurch did not send in his bill for attending Grace until 13th January 1680, and the charge was 7s which included the cost of the physic. Hooke paid the bill on 19th January.

Grace was more fortunate than Tom Gyles, Hooke's nephew, who died of smallpox. The figures given by Dr Kenneth Dewhurst in his book *John Locke, Philosopher Physician, Philosopher*, show that in a population of London of about half a million, there were 16,000 deaths in 1667, and 17,000 in the following year, and that smallpox was the cause of death of 1,196 and 1,468, respectively.

Tom Gyles had received different treatment. On 8th September 1677 he had complained of a crook in his back, and Hooke wrote to his brother John and to R. Gyles, Tom's father, to take him home. On the following day Tom roared with pain in his back, and Dr Diodati judged that Tom had measles. Tom was very ill the next day. Gidly came on the 11th and refused to let blood. Tom's throat was very sore, and he almost choked. Dr Diodati made another visit, and as Tom had pissed blood all day, he advised that Tom be let blood at the nose and mouth. At 8 o'clock Mr Gidly and Mr Whitchurch came and let 7 ounces of blood from the arm and under the tongue, but in spite of this, Tom continued bleeding from his throat and nose all night, and passing blood.

Hooke rose at five in the morning and called on Gidly, and he sent him to the doctors. Hooke spoke to old Dr King who affirmed that pissing blood in smallpox was mortal. These two and Dr Mabletoft called to see Tom at nine, and concluded that Tom was irrecoverable. Tom spoke very piously, began to grow cold, to want covering and to have little convulsive movements, and after falling into a slumber seemed a little refreshed and spoke very sensibly but, composing himself again for a slumber, he rattled in his throat and died. It was about fourteen minutes after noon; he seemed to go away in a slumber without convulsions.

'I gave Mr Gidly 5s and the searchers 16d. (These were the people responsible for ascertaining the cause of death, and reporting to the parish authorities.) I drank half a pint of sack.'

Tom was buried in All Hallow's churchyard, and his funeral was attended by at least 50 people. On 25th October Hooke received a letter of gratefulness from R. Gyles for his kindness to poor Tom.

Dr Sydenham's treatment for patients suffering from smallpox was quite different to that given to Grace and Tom. He bled his patients, but more importantly applied cool soothing lotions to the spots, gave plenty of liquids to avoid dehydration, and sometimes an emetic. His general principle was to cool the patient, realising that nature often brings about its own cure with the help of simple remedies. Most of his patients survived by his forms of treatment.

The Dr Millington mentioned in the extract from Hooke's diary was also an Old Westminsterian, one of the founding Fellows of The Royal Society and later President of The Royal College of Physicians. He was in favour of Dr Sydenham's cooling methods for fevers, and dismissed many of the orthodox methods then in use. John Locke also advocated Dr Sydenham's methods.

Grace continued as Hooke's housekeeper until her death

in 1687. From an observation made by John Aubrey, that she had a dark mark on one of her breasts, she may have died of cancer, but this is only conjecture. Her death was a great loss to Hooke, and as there is a gap between the dates of the two diaries, there are no records of his feelings at that time.

The entries in the second diary which started on All Saints Day 1688 are written in a similar style to that of the first, with brief laconic statements. There are no x signs, so he may have lost interest in that after Grace's death. Of his two housekeepers mentioned, Martha was often criticised, once for breaking a dish, for being out until 12 o'clock, being insolent, proud and surly, His other maid Mary did not come in for such criticism and on 20th May 1693, Hooke gave Mary Grace's hood, so he thought more highly of her.

DISH OF THE DAY

5th Sept. 1672	"Drank ale, eat eggs and milk, slept very little."
6th Sept. 1672	"Eat raw milk and bread, slept very well."
17th Sept. 1672	"Eat roasted chicken, took I tobacco, slept pretty well."
10th Oct. 1672	"Eat china broth and mutton broth. Slept pretty well."
19th Oct. 1672	"Slept very well after eating herring, broth, and aniseed water."
26th Oct. 1672	"About 6 at night I eat chicken broth and milk thickened with eggs and took conserve of Roses and Dr Goddards Syrupe of poppys. Slept disturbedly, but no griping."
9th Nov. 1672	"Eat roast lamb and chicken broth, with conserve of roses. Slept not well, rose at 6."
12th Nov. 1672	"Eat meat and apples at Rose, Fleet Bridge."
26th Nov. 1672	"Eat meat and broth, slept ill."
25th Dec. 1672	"Upon eating broth, very giddy. Eat plumb broth, went very well to bed but slept but little."
1st Jan. 1673	"Much disturbed with giddiness in my eyes and head after drinking new ale and eating mutton pottage, slept a little after it. Better after eating pullets and drinking ale."
14th Jan. 1673	"Eat Beef at noon, Eggs at night, Slept not well."
15th Jan. 1673	"Refresht after washing hands, face, eating chicken and drinking beer."

18th Jan. 1673	"D.H. Eat neats feet. Walk."
27th Jan. 1673	"Eat stewd mutton with lemon agreed well eat gruel and chick."
28th Jan. 1673	"Veal dinner rose in my stomach as salt as brine."
6th Feb. 1673	"Eat green ginger from Sir A.G. King."
15th Feb. 1673	"D.H. cold mutton. somewhat vertiginious – Slept a little after dinner."
23rd Feb. 1673	"At Mr Hills. Eat barley broth. Slept very disturbedly."
4th Mar. 1673	"Eat broth with great stomach began to smell some good sents and some bad. Slept very well."
2nd Apr. 1673	"Eat boyld veal and drank ale. Slept well and was very hungry next morn."
3rd Apr. 1673	"D.H. on rost veal."
15th Apr. 1673	"Eat Stewd pruines with senna which agreed well."
10th Jun. 1673	"Eat pease drank Dr Goddard's ale. ℞ Slept well."
1st Aug. 1673	"Garways. eating salt fish and Drinking much wine I found myself much better."
9th Aug. 1673	"Eat beans and bacon going to bed. ℞ slept but 2 houres."
2nd Sept. 1673	"In the evening to Garway and Lord Brounker. Returnd slept not well after eating rice pudding."
13th Sept. 1673	"Eat apple pye and slept very well, rose refresht."
1st Nov. 1673	"All this week I found myelf much better by eating fish suppers and drinking Madam Tillotsons Ale."
2nd Nov. 1673	"Supd on mutton which agreed not. took aldersgate, much out of order next morning."
3rd Jan. 1674	"At Garways till 10. Eat 5 larks. Letter from Brother. took senna."

13th Jan. 1674	"D.H. on Neatsfoot and onions. At Garways. Bespoke dancing shoes."
27th Jan. 1674	"At Garways. Discoursed Collins, Ludwick & c. Mr Wild, Mr Aubrey, Mr Hill, Eat beef, Rost apples, water. Slept pretty well."
28th Jan. 1674	"Eat fryd beef and rost apples."
12th Apr. 1674	"Cox here. Dind at his house on pork and veale."
21st Apr. 1674	"At Garways. Eat pullet and cheese cakes, drank a little new beer. Vomited, after which, I was much refreshed."
29th July 1674	"D.H. Mutton and Carrots."
20th Aug. 1674	"Nell dind here on fish."
1st Oct. 1674	"Eat Oysters with Sir G. Makerman. At Garways in little room."
1st Nov. 1674	"Dind at Sir A. King on beef and goos."
7th Dec. 1674	"D.H.on beef and small beer. Agreed well."
18th Dec. 1674	"Mary receivd Honey and cheese. Both very good."
29th Dec. 1674	"D.H. Goos Py. Nell cryd."
30th Jan. 1675	"D.H. mutton. Took senna. wrought."
11th Feb. 1675	"Saw Sir Christopher Wren. Bought in Fleet Street for 4s mellons."
8th Mar. 1675	"Eat a lump of Honey in the Morn and fasted after it."
15th May 1675	"Eat brothers crab."
8th Jun. 1675	"Dind with Mr Boyle on pease."
24th Jun. 1675	"Fryd Lamb."
9th July 1675	"At Garways. Whistler, Flamstead, Tompion, Hill, & c. Eat Cream agreed well. – Sweat."
20th July 1675	"Dind with Sir Ch. Wren on beans & c."
23rd July 1675	"D.H. Bacon and Pullet. Slept at Haaks. Lent Tompion IOs."
11th Aug. 1675	"D.H. Rabbit."
18th Aug. 1675	"At Garways with Fitch and Hayward.

	Vomit from Hewk. Eat cheese cakes and apple. Slept well and agreed well."
6th Sept. 1675	"At Storeys cold venison and codling."
7th Sept. 1675	"D.H. Removed Coles in celler upon full stomack of Beef and Cabbage. Mightily refresht."
10th Sept. 1675	"At Dr Allen eat fruit and long plums agreed well."
8th Oct. 1675	"Dind with Waters in Shooe Lane on beef and cabbage."
11th Oct. 1675	"Bates set up work bench in turret, he would take nothing I gave him a Barrell of oysters and a bottle of Canary."
2nd Oct. 1675	"Eat Pidgeons. Slept well. *Deo Gratias*."
12th Oct. 1675	"D.H. Scarborough, 2 pidgeons. Milk made me sick. Tompion here. Eat Oyster and fresh herring that Scoured me."
18th Oct. 1675	D.H. Eat oysters and polisht 25 foot glasse reflecting. Completed Theory of fly watch. Was very well till eating pidgeon pye. Played cheese with Mr Haak."
31st Oct 1675	"Roast beef agreed not."
2nd Nov. 1675	"Dind on Graces goos. Mary forgot the filling."
3rd Nov. 1675	"Eat goos and apple. Agreed well."
4th Nov. 1675	"D.H. Very sick after eating marrow and rosted pidgeons."
13th Nov. 1675	"Eat hen and duck. – tooke aldersgate 3 spoonfulls."
14th Nov. 1675	"Eat Pidgeon broth and pidgeons, wrought thrice. Eat preserved walnuts. – disturbed stomach by taking strong water."
26th Nov. 1675	"Dined at Shoe Lane 7d."
13th Dec. 1675	"Eat Gease feet."
27th Dec. 1675	"Rose in the night eat turkey at 4. Slept a little after."
2nd Jan. 1676	"Could not sleep after Chocolat till eating salt beef at 2 in the morn. Slept well. ☿."

11th Jan. 1676	"Dind with Haak on 3 cocks. chesse."
19th Jan. 1676	"D.H. Neats tongue and cabbage – fit of colic."
22nd Jan. 1676	"Eat sturgeon."
28th Jan. 1676	"Eat rost beef which agreed extremely well. Slept well."
29th Jan. 1676	"By coach home 1s. Eat sturgeon vinegar and sugar. Agreed not."
9th Feb. 1676	"Haak invited me to Dinner of fish."
17th Mar. 1676	"D.H. Pease and marrow."
18th Mar. 1676	"Eat oysters which scowred me and pease porredg."
20th Mar. 1676	"D.H. pease and veal. With Mr Haak."
23rd Mar. 1676	"Eat calves head and pease. Slept feverish."
3rd Apr. 1676	"Drank port and eat pease porridg."
21st Apr. 1676	"D.H. on chicken. Fitted up Graces closet."
22nd Apr. 1676	"Eat eggs and drank wormwood beer at Nell's."
28th Apr. 1676	"Eat chicken and drank white wine. Slept well."
3rd May 1676	"Eat eggs, vinegar and stewed prunes. Slept well."
5th May 1676	"Eat several apples."
8th May 1676	"Mary boiled not cabbidge. Very ill in stomach."
11th May 1676	"Eat Hen and Sprouts agreed exceeding well."
4th Nov. 1676	"Dind with Dr Whistler on pease porridge."
30th Dec. 1676	"Eat parsnips with salt beef. Drank claret. Eat milk porrage in the morn."
1st Jan. 1677	"Eat beef and parsnips, slept well. Lent Grace little play book."
4th Feb. 1677	"Eat Goos not well rosted made me very sick."
27th Jan. 1678	"D.H. Beef and pease. Very melancholy all things crosse. *vindica me Deus.*"

4th July 1678   "Dind with Sir Ch. Wren on beans and bacon."
20th July 1678   "Eat tart with Boyle, mett Bennet there."
7th Aug. 1678   "D.H. Poysond by Drinking glass of sour beer."

On 24th February 1674, Hooke had been told by Mr Whitchurch of the extreme deliciousness of the queen pineapple. Three years later at a meeting with Sir C. Wren and Sir J. Hoskins at Childs, he was told that Wren had eaten a queen pine English growth like a Hartichock. The pineapple had first become known in England after the capture of Jamaica, and the first plant to be grown here successfully, and bear fruit was at Dorney Court, where the local pub commemorates this fact. Hooke was able to taste one himself on 2nd August, and it cost him 2s. He bought 12 oranges from Mrs Davys on 30th August 1678, which cost him 6d and he ate 7 of them which agreed well. He went in Lord and Lady Ranelaugh's coach pulled by six horses, to Chesweck on 14th June 1678, and ate many strawberries, but no mention of the cost. He ate 2 lbs of cherries at home on the 21st July 1677, but did not state the cost.

A satirical book published in 1664 under the title *Mrs Cromwell's Cookery Book* was also notable for the lack of fresh green vegetables in the diet of that time, as was a book published more recently about Mrs Pepys' skill as a cook. In Hooke's menus given here there is one mention of sprouts and a few of cabbage, but the meat ingredient seems to be the most important.

BROTHER JOHN HOOKE

Hooke's brother John was five years older than him, and at the time that the first diary was written, was a grocer at Newport, Isle of Wight. The hearth tax returns for the years 1664–1674 show that a Mr John Hooke was resident in the High Street, Newport. Some of the first entries in the diary relating to John show his need for the new issue of farthings for his business.

On 22nd August 1672, "New farthings first made publique", and on the same day, "Received of Dr Wren £50 in new half-crowns". In November of that year he "Received 25 shillings in farthings from Hore, sent to Brother John".

For many years there had been a shortage of small change for anyone in trade. According to references in *Coin Collecting* by Milne, Sutherland and Thompson, Stuart and Commonwealth governments had neglected to mint coins of small denominations for general use. The book also states that James I had authorised Lord Harrington to mint farthings, but people were suspicious of them, and most of them found their way to rubbish dumps. Local tradesmen met the need for coins of small denomination by minting their own tokens. These were only acceptable in a small area, but where there were many shops, as in the Strand, there were 31 of them issuing tokens, and shops may have accepted other tokens and then settled up with the issuer.

Tokens issued by John Hooke are listed in a book devoted to tokens of Hampshire. The inscription states that the token of John Hooke, Newport, Grocer, is not dated and does not have the conventional sign of a grocer, either the nine cloves, a camel with a sack on its back, or three sugar loaves, as was the sign of Edmond Castle, a grocer of Burford.

Charles II had authorised the issue of official farthings and halfpennies, and the withdrawal of tokens, so John

Hooke was obviously complying with this. The new farthings are of interest, as on the obverse side they show the figure of Britannia, modelled on the figure of Frances Stuart, Duchess of Richmond, a beautiful lady who resisted the advances of Charles II.

When John Hooke sent the 20s in payment for the farthings, he also sent up a basket of grapes. On 21st December 1672, John Hooke sent a goose, one hare and 20 poultry to Robert, and from time to time similar gifts were sent from the country.

When his brother sent his account to Robert on 15th February 1673, he sent two fowls with it. Robert bought a bed from John on 6th January 1673 at 10d a pound, the total cost being 54s plus carriage. Robert later received a pot of honey containing 5 quarts for which he was charged 15s, although his brother had first told him the cost was 6d per pound.

John Hooke's grocery business does not seem to have thrived as from time to time there are requests for loans of £20 or £40. In August 1675 John wrote to Robert about purchasing an estate at Avington. Robert replied that he thought that the price was too dear, but, in February 1676, he changed his mind and offered to pay £4000 for the property, stating, "I was rather for a life to doe good to friends while I live."

According to the *Victorian County History for Hampshire*, Vol V. p. 229, the family of Leigh owned the manor of Avington, or Carisbrooke, and that an unenclosed part of Parkhurst Forest was sometimes known as Avington Park. In April 1676 Robert had changed his mind again, and put off buying the property.

Robert Hooke sent a request for his brother to send him some shells and petrified mud, and sent him a "chymicall book", which he had bought at Martins for 8s 6d, and a silver tobacco box which he had bought from a goldsmith at the Sign of the Rose, at Holbourne Bridge at a cost of 20s. He also sent him a perspective glass (a telescope) and notes on the proceedings of Parliament.

John Hooke was gradually getting into debt with Robert, and in December 1676 when Robert was making up his accounts for the year he noted, "Due from Brother J. Hooke upon bond for interest for £200 for four years £50". In March 1677 he sent John £50, and had a further request for £20 or £40 in June 1677, and in response to these, sent him £20 on 11th July 1677.

On 27th February 1678, Hooke recorded in his diary, "This morning Brother John Hooke died".

Three days later, after a discussion with Sir John Hoskins, he "Spake to the King for brother Hooke's estate, but the King replied that Sir R. Holmes had already begged it for his wife and child."

Sir Robert Holmes was Governor of the Isle of Wight, and had been captain of the ship in which Hooke had sailed in 1662 to test the suitability of using clocks at sea for determining longitude.

The biographical note in volume X of Pepy's Diary states that Holmes used his office of Governor to turn it into a lucrative satrapy, but the article also mentions that Richard Ollard's biography shows that he was a superb man of war.

A SUMMARY OF SOME DETAILS OF HOOKE'S LIFE, FROM THE DIARY

There are many references to Theodore Haak, a German scholar and influential thinker among the first Fellows of the Royal Society, having been elected on 4th December 1661. Many diary entries read, "Haak chess", and on one occasion, "Beat Haak at chess". Haak translated letters from Hevelius and Leeuwenhoek from Dutch for Hooke, until he had become sufficiently proficient in that language, for which he had lessons from Blackburn. Hooke noted that Haak broke his nose by a fall at Jonathon's. They often smoked at their meetings, Hooke noticing that Haak had a meerschaum pipe. Hooke sometimes smoked two or three pipes and felt much refreshed. He also gave up the habit from time to time, and felt much better. He bought a quarter of a pound Virginia on 13th March 1677, which cost him 9d. A metal tobacco box cost him 5s, and 2s to fill it.

There are 154 coffee houses listed in the appendix of the diary published in 1935, as being in the City and its suburbs. The two most frequently visited by Hooke were Jonathon's and Garraway's, both were in Change Alley, conveniently near to Gresham College to be visited after the meetings of the Royal Society. Hooke mentioned the rude flirts at Garraway's, and that the Lord Mayor had forbidden him to go there. He washed his feet there once, and noted the new singer on another occasion. He was present when an elephant was sold for £1,500, and mentions the Hudson Bay Company having a meeting there, on 4th August 1679, when Sir J. Hoskins, Sir J. Norberry, Sir J. Hayes were present. There had been a meeting of this company on 2nd July, when Sir Christopher Wren, Hill,

Bent, Colwell and Evelyn were present, but from the entry it is not clear who were members of the Hudson's Bay Company.

When Garraway's daughter died Hooke designed a monument for her, which was made by a stonemason called Bomstead, and erected in St Peter Poer church in Old Broad Street. This church had become ruinous by 1788, and the brasses were sold to a plumber in the Minories. The church was finally demolished in 1907. Garraway showed his particular treasure to Hooke, a collection of agate cups.

He was a brisk walker, and enjoyed walking in the fields north of the City. He recorded running up and down in the gallery at Gresham College, until he sweated, and playing ball on two occasions. He bought some dancing shoes, and informed Wren of this, but Wren did not approve, and there is no mention of Hooke trying them out at a dance. He bought a sword and belt, and a pair of French pistols. It was customary for gentlemen to wear swords, and use them. Some duels were noted by Hooke. He had once scuffled with a rogue, and broke his shins, so these weapons may have been for his protection.

Twice he was nearly run over by a coach, and when Mr Povey said he would prosecute the owner, Hooke declined the offer. He noted the times when he travelled in Wren's coach, but unlike Pepys, did not buy one of his own. He carried out experiments in the design and springing of coaches.

Like Pepys, he generally rose early, perhaps to save candles, and to work in daylight and prevent strain to his eyes. Pepys was particularly worried about this, and Hooke later found spectacles useful. Both men would have been used to early rising at their respective schools and, being very active men, the habit remained. Sometimes Hooke would work all through the night, and then have a nap after dinner. Once after dining with Tompion he slept in his hammock and fell out.

His opinion of astrology was that it was vain. His

patron, Robert Boyle and his fellow Westminsterian John Locke, although being followers of the scientific methods advocated by Francis Bacon, took the advice of astrologers before planting their peonies. Charles II took the advice of astrologers when at the races at Newmarket, as to which horses to back. These two examples are given by Christopher Hill in his book, *The English Bible and the Seventeenth Century Revolution.*

As well as drinking a variety of waters, Dulwich, Epsom, Tunbridge and Tattenham Court, he drank brandy, port, claret, sack, and birch juice wine which he found to be delicious. He also had a barrel of Flamsteads ale, and Tillotson's ale. There are a few instances when he recorded that he had been drunk, and one occasion, on 12th June 1675, when, "Drunk. promised to pay whatever I signed to be paid."

On 16th September 1675, when at Garraway's with Dr Whistler and Lem, he noticed that Sir J. More was fuddled horribly.

He was a gregarious person, who liked to meet people, particularly those who had travelled abroad, such as Captain Knox, who had been a captive in Ceylon for twenty years. Hooke designed a sextant for Francis Vernon, another experienced traveller and old Westminsterian. This sextant was not finished before Vernon set out on what was to be his last journey, and Hooke had said he would send it on to him. On 28th May 1677 Hooke heard from Dr Mabletoft, another old Westminsterian, "the ill news of poor Fr. Vernon's death, killed by Turks with cimeters at Ispahan".

Hooke heard of news of the North West passage ship on 13th December 1679, an expedition which Evelyn noted in his diary. He had dined at the Admiralty with Secretary Pepys, and supped at the Lord Chamberlain's. "Here was Captain Baker, who had been lately on the attempt of the North-west passage. He reported prodigious depth of ice, blue as a sapphire, and as transparent. The thick mists were their chief impediment, and cause of their return."

On 8th January 1674 he saw Oldenburg's book of Finland to Lapland. "N.B. Their way of sliding on snow." He was interested in the account of the natives in India climbing palm trees barefooted, and he asked travellers going to that country to send him details about it. He noted the report that ships loading in Greenland could be to a different level than in England. There are many references to earthquakes in Teneriffe, India, and other parts of the known world, all of which were mentioned in his lectures on earthquakes. He noted what he thought was a nonsensical hypothesis of Flamstead of earthquakes in the air at a Royal Society meeting on 3rd May 1689.

Items concerning the supposed Popish Plot, and Titus Oates are mentioned and the trials of those involved. The Lords' treason trial is mentioned on 4th December 1678, and that five priests and Langhorn were condemned on 13th June 1679. The punishments of the guilty were noted, as 17th December 1675, when he saw eleven pirates hanged at the Old Bailey.

Sir Henry Blount (1602–1682) whose life is described by John Aubrey, thought that his servants would do better to go to see the executions at Tyburn, as this sight would have a more beneficial effect on them than all the sermons. Hooke noted the whores slashed at Bridewell, and Muggleton in the pillory before the Temple, where he was pelted with rotten eggs on 23rd January 1677.

On 13th September 1680, he saw "Celier condemned £1,000 and pillory". This was Elizabeth Cellier who was fined and pilloried for libel. There are no views expressed about these punishments, so they were probably accepted as the normal thing at that date.

Hooke's request not to have to perform any more operations on dogs would suggest that he was compassionate towards animals, as also his noting that Blackburn kicked a dog, as though he disapproved. His obvious concern shown when first Tom Gyles and then Grace caught smallpox certainly is evidence of his feelings for fellow human beings.

At the Restoration, the date of King Charles' I execution

and the date of Charles II return to the throne were ordered to be kept as fasts, and Hooke notes these in his diary. He fasted on other occasions, and once just fed on honey.

He noted some sermons, such as Archbishop Parker's and prayer on 30th January 1677 when there was much discourse about it afterwards at Garraway's with Mr Hill and Mr Erskine. He heard Dr Barrow preach at the Charterhouse on 17th August 1673, and Dr Burnet preaching about spirits and against the Pope on 8th December 1678. He heard the same preacher on the subject of peace on 22nd February 1680 at the Rolls. A sermon on 20th September 1676 given by Dean Tillotson about the plain man's guide to Heaven was discussed with Sir John Hoskins at Child's coffee house. Another sermon of which Hooke was informed was that against Mr Hobbes' new book of philosophy. He was not as meticulous as John Evelyn, who noted all the sermons in his diary. Nor is there the detail in his brief jottings to show whether he took more interest in the ladies in church, as Pepys did, or kept his mind on the sermon.

Hooke seems to have continued with some country pursuits in the yard at Gresham College. On 2nd August 1673, he searched for the lost beehive, and on 5th January 1675 he paid Mr Colderwand for sending home beehive 2s 6d.

On 2nd March 1674 he noted, "Mr. Hoskins 12 scarlet peas", but it is not clear from the entry whether Hooke was buying them. He bought 18 flowerpots on 9th February 1677, and some at an earlier date, so he may have had a small garden. He bought 3 bush roses for 3d on 6th July 1675, and on the 8th picked some.

The yard was used for dumping his night soil, and this was cleared from time to time by Wadlow.

February 20th 1673, "Wadlow promised to clear my yard", which he did on the 24th. "Wadlow's men removed dung out of yard."

It was probably taken to the laystalls at Dowgate, and then loaded on to barges to be carried further down the river and dropped overboard.

His coals were dumped in the yard, and on 20th December 1672 he paid Mar. £3 14s for 2 chaldron of coals. The footnote states that a chaldron was about 27 cwt.

The price was the same on 10th June 1673, and on 15th August 1673, he paid Berry 37s per chaldron. He paid Mrs Hearn £2 12s for the same quantity on 14th September 1674, and on 21st July 1675, ordered 2 chaldron from Hearne for 26s 6d. He received 2 loads of coal from Fitch on 7th September 1675, but did not state the price. He received a fine chaldron and a quarter of coles from Mr Hammond on 12th September 1677, and paid 5s for shooting them, there being in all 63¾ sacks, and on 8th October he paid Mr Hammond himself at his own house £5 for 5 chaldron of "coles", Mr Crawley witness.

In December 1666, Pepys had complained of the high price of coal, £3 3s per chaldron, a great want of them in the City, due to a fleet of 200 Dutch ships and 14 Dutch men-of-war being between the colliers and the Thames.

A problem with burning so much coal was the smoke produced. On 28th September 1676 Hooke rode with Sir J. Hoskins to Banstead, and was hurt in the testicles by the jolting horse. He observed a deep wall echo, and also the cloud of smoke over London half a mile high and above twenty miles long.

Three years later, on 17th November 1679, he noted, "A great fogg". Evelyn noted a similar dense fog on 15th December 1671, when, "It was the thickest and darkest fog on the Thames that was ever known in the memory, and I happened to be in the midst of it.".

He reported another on 15th November 1699, that was so thick that people lost their way in the streets. He had presented his paper *Fumifugium* on 13th September 1661, which was a criticism of London's foggy atmosphere, and advocated the planting of trees to lessen its effect. The Clean Air Acts of this last half century have produced the improvement that he wanted.

The City had imposed a tax of 1s per cwt on coal landed

at London, and on 6th April 1677, when Hooke was with Sir John Hoskins at Garraways, Hooke asked his opinion about the order authorising the payment of his salary as surveyor of new building, of £50, which was to be paid out of the money raised by this tax.

Once a fire broke out in his stable, "which was like to be very Dangerous but it pleasd God to prevent it. Blessed be his name".

Twice he was nearly run over by a coach, once on 17th July 1674, and again on 15th September 1674 when, "Almost run through by the pole of a coach but escaped Deo Gratias in Mr Povey's coach."

He attended plays, as on 6th December 1673, when he saw *The Empress of Morocco* at the Dukes Theatre 1s 6d. The Duchess was there. He went with Sir Christopher Wren to see *The Tempest* at The Playhouse on 20th June 1674, paid 3s. He went to see *The Virtuoso* on 2nd June in the company of Tompion and Godfrey. This play by Shadwell poked fun at Fellows of The Royal Society and their experiments, and especially Hooke, who was recognised by some in the audience.

"Damned Doggs. *Vindica me Deus.*" was the comment in his diary. It was an attempt to belittle his work, and may have been repeated since for the same purpose. On the following day he met Sir J. More, Flamstead and Hill, who had just come from watching the play and "Floutingly smiled". Hooke seems to have seen the play again on 1st July, and on 3rd July he bought a copy of the play from Mr Martin for 1s.

At the time there were 200 coaches licensed to ply for hire in London. On 21st November 1672 Hooke hired a coach from Arundel Library to Gresham College which cost 1s. After visiting Sir Christopher Wren on 23rd May 1674, Hooke took the coach back to Gresham College, which again cost 1s. The charge was the same after visiting Mr Montagu on 10th February 1675. When he went with Davys on the Highgate coach on 18th August 1675 it cost him 4s. When going by water to Queenhithe on 25th May

1676, the cost was 6d. He went with Sir J. Hoskins and Lady to Greenwich castle on 15th September 1676, and saw Halley and Linger there. They saw the Queen's House, which cost 6d, and the cost of the coach was 1s.

On 3rd October 1676, he visited Sir Christopher Wren where they were busy passing Peirces and Marshall's bills till seven. "By coach 1s 3½d". Hooke travelled by water to the Paules on 2nd October 1679, which cost 6d, and a journey to the Stilyard on 9th October, cost him the same. Most of his business within the City was done on foot.

As Waller described him, "He went stooping and very fast (till his weakness a few years before his Death hindred him)."

At the time of the first diary Hooke was between 37 and 45 years of age, and Waller did not have much contact with him until he was accepted as a Fellow of the Royal Society in 1681.

Hooke was a capable horse rider, as was shown on his visit to Welling on 4th May 1680, and his return on 5th May, when he rode to Dunstable by 7.30 in the morning, thence by coach to London at 6 p.m. An even longer ride was on his return from visiting Lord Conway, when "after laying very scurvily at Islip", he took horse at four and arrived at Beckonsfield by ten. He dined there and arrived in London and Gresham College by six. Davys, his companion, was "seased with his ague, I was not in the least weary. Went with Society to Jonathons stayd with them till 10 at night. Slept well. Soli Deo Gloria."

He frequently went for walks into the fields north of Moorgate and on two occasions played ball. He fasted on a number of occasions and generally on the anniversary of the death of Charles I.

There is only one instance of him taking the Sacrament, at St Peters Poer on Sunday 4th May 1673, and two days later he was sworn before Sir W. Wild at Westminster, together with his fellow communicants to comply with the law relating to church attendance in force then. Once he noted, "Extraordinary much bellringing" and, "Saw mock

Pope carryd in procession". On 5th November 1675, and on 29th May 1677, he noted "bonfires for King's Restoration". But these dates were not noted annually for such celebrations.

Inspite of having had ill health in his childhood he was capable of sustained work, and at times noted his good health, as on 10th October 1678 when he wrote, "Exceedingly well in health" and again on 17th November that year, "exceeding well in health Deo Gratias".

Hooke's Purchases of Tools

17th Mar. 1673	"Bought of Mr Marquill Lathe for 2 Guineys."
18th Mar. 1673	"Harry cleansd lathe. began wheel cutting engine."
16th Sept. 1673	"Bought sliding candle stick at the 5 Bells in Lothbury for 4s. paid. Harry made microscope tooles of 2/10 of an inch."
28th Jun. 1675	"At Tompions he mended old watch, with three circulating balance. Bought 2 Rasps, 2s. Lamb bought oylstone, 18d."
11th Aug. 1675	"Bought of ironmonger 33 tiles 5s, wireplyers 6d, Bicorne 2s 6d, Firehasts 20s 10d, Wire turne bench 1s."
11th Oct. 1675	"Bates set up bench in turret, he would take nothing, I gave him a barrell of oysters and a bottle of canary."
13th Oct. 1675	"Received from Bates two small plaines."
22nd Oct. 1675	"Gave Davys 3s for 3 pair of hinges."
8th Nov. 1675	"Bought at Hayes in Wood Street 12 chistles and paring chisall 4s, 3 pairs hinges 1s, nayles 4s, in all 9s."
13th Nov. 1675	"Paid Davys 6s 6d, for plow 2s 4d, 2 plain irons 8d, saw 3s 6d."
10th Nov. 1675	"D.H. With Davys bought Peirces bits 4 small at 3d, larger at 4d, 1 half inch 1s,

	nimble stock 6d."
11th Sept. 1676	"Davys brought home 3 8 foot tubes. Caried turning wheel into turret."
12th Sept. 1676	"I paid Davys at Man's coffee house for 2.8 foot tubes 5s."
16th Sept. 1676	"Set up the mandrill."
18th Sept. 1676	"Bought of – brasse plate at 1s 6d per lb. 2s 9d paid. cutt it into instrument joynt. Davys sent in rules too thin."
26th Jan. 1677	"Barton sent home iron crank for turning wheel."
30th Jan. 1677	"Took out of Scarborough trunk, an anvill and 2 hammers contrived the twisting spring."
2nd Feb. 1677	"I payd Scarborough for shovell 1s 6d, pitaxe 32d, trowell 12d."
7th May 1677	"Fitted turning wheel."
6th June 1677	"At Hayes, ironmonger, about locks."
7th June 1677	"Sent for locks to Hayes. To Mr Harveys about locks."
17th Aug. 1677	"Bought brasse wheels & c. 1s 6d, 8 drills, a hammer and pincers, 2s paid."
21st Aug. 1677	"Bespoke of Walker file-handle."
22nd Aug. 1677	"Through Morefields, from Smith, Tooth file frame."
23rd Aug. 1677	"Crawley bought 4 files, 1s paid."
24th Aug. 1677	"To Mr Boyles, dind with him, begged long screw quadrant, and took mandrill. Directed Laboratory."
25th Aug. 1677	"Sent Tom for tooles to Boyle and Slayer, he brought the long screw instrument, a lath with two poppets and a screw mandrill, a tool to cut wood screws."
22nd Sept. 1677	"Took from Mr Boyle his wooden frame, and one poppet head for Lath.
8th Oct. 1677	"To Sir J. Hoskins, his fire shovell and tongs made of iron 5s, his fire hooks and fork 2s, Iron back grate, grate back, & c.

	25s."
14th May 1678	"Bought of brasier Cateaten Street 2s of Brasse."
28th May 1678	"Paid Coffin 2s 6d for table and hinges letting in."
31st may 1678	"Through minorys, pistoll for cylinder, 9d."
25th May 1678	"D.H. Slept p.p. fitted bow to lathe."

These are details not mentioned in either Evelyn's or Pepys' diary. Pepys gave the cost of timber, rope and other ship building materials, but not of the tools used. Neither are these details mentioned in Trevelyan's *Social History of England*.

ROBERT HOOKE
THE LEONARDO
OF LONDON

HOOKE'S *MICROGRAPHIA*

Hooke's first major publication, *Micrographia*, is a summary of most of the scientific work he achieved before the age of thirty.

In the opinion of R.S. Westfall, the American historian of science, "*Micrographia* remains one of the masterpieces of seventeenth century science. Like Galileo's *Nuncius*, *Micrographia* presented not a systematic investigation of any one question, but a bouquet of observations with courses from the mineral, animal and vegetable kingdoms. Above all, the book suggested what the microscope could do for biological science."

This brief appreciation does not include the full scope of the book, which commenced with Hooke's philosophy of science, and the necessity to aid the human senses in the understanding of the environment in which we live by inventing instruments to increase our awareness of its detail, and our appreciation of the omnipotence of its Creator.

Two illustrations from the book have been used in recent years. One was of the snowflake, used on a postage stamp commemorating the 150th anniversary of the Royal Microscopical Society. The other has recently been used in an advertisement in London buses to bring the public's attention to pests who try to avoid paying their fares. This is the picture of the flea which, unbeknown to science at that time, was the insect responsible for spreading the Plague.

The preface of the book describes the instruments which Hooke designed, and his reasons for making them. There is an illustration of his compound microscope, and a description of its fault in giving chromatic aberration to objects viewed through it. He found that using a single lens greatly reduced this effect. He obtained a magnification of thirty using the compound microscope, and of eighty when using a single lens, as when examining silkworms' eggs. He made a spirit of wine thermometer (alcohol), and explained how he graduated this. This was a sealed thermometer to avoid any change due to changes in atmospheric pressure and marked its zero at the alcohol level when water started to freeze. Later he extended its range below zero by using a freezing mixture of salt petre and common salt. He devised the wheel barometer, by which readings of one hundredth of an inch could be made.

A device for grinding and polishing lenses was among the objects illustrated.

The experiment by which he had demonstrated to the Royal Society how liquids rise to different heights in glass tubes of different internal diameter was one of the illustrations. When he gave this lecture he stated that this might explain the rise of sap in plants, and of oil in the wick of a lamp.

He began his microscopical observations, as Euclid did his geometry, starting from a point; the tip of a needle, a line, the edge of a razor, an ink blot. His opinion of the first two man-made objects was that "these were only some of the many signs of the rudeness and bungling of Art, whereas in the works of Nature, the deepest discoveries show in the greatest excellencies, an evident argument that He that was the Author of all these things was no other than Omnipotent."

A comparison of the crudeness of the point of a needle with that of a bee's sting will show the force of his argument. His comments on the magnified ink blot were that it might be the way of conveying secret information, a prediction which has been put into practice.

He mentioned that he had seen some minute writing, where the Lord's Prayer, the Ten Commandments, together with half a dozen verses from the Bible had been written on a piece of paper only as wide as a two penny piece. He may have been referring to the work of Edward Cocker, the noted engraver, who had engraved the scale on a ruler for

Samuel Pepys, and was famed for the minuteness and exactness of his work.

Hooke examined some silken flax and fine washed silk, and he wondered whether it would be possible to spin a kind of artificial silk out of some glutinous substances. With the advances in chemistry, made during the next two hundred years, Chardonet prepared rayon in 1884.

Using a single lens he was able to observe the action of the hard point of a stinging nettle when it penetrated the skin, and see a small amount of liquid be forced into the wound. This made him, "wonder the Invention of that Diabolical practice of poysoning the tips of their Arrows and Poynards might have received their first instance in natural contrivances".

He wondered whether the experiments of Dr Wren in injecting liquors into the veins of animals, might be extended to humans for curing "several humane distempers such as Gout, Dropsie Stone etc."

He examined Muscovy glass (mica), and noticed that the colours seen varied as the layers were pressed together.

"With the Microscope I could perceive that the colours were arranged in rings that encompass the white speck or flaw, and were round, or irregular, according to the shape of the spot which they terminated; and the position of the colours, in respect to one another, was the same as in the Rainbow. The consecution of the colours from the middle of the spot outwards being Blew, Purple, Scarlet, Yellow, Green."

He also observed that these coloured patterns were to be seen on the layers of thin glass, when he had blown it and on the soap bubbles blown by children. Similar colours were seen on the surface of hardened steel, which he thought were caused by a thin surface layer produced by heat when it was tempered.

Isaac Newton acknowledged that his interest in this phenomenon had been aroused by Hooke's observations published in *Micrographia*, and his own experiment, "Newton's Rings", showed the regularity of the spacing of the colours.

From Newton's measurements, Thomas Young, in 1806, was able to calculate the wavelength of red light as one thirty-six thousandth of an inch, and that of extreme violet as one sixty thousandth of an inch. Hooke was of the opinion that light travelled in wavelike pulses, the pulse length of each colour being distinctive to that colour. Newton thought that his experimental evidence was more strongly in favour of a particle theory. By his experiment of Young's Fringes, Young was able to show that light travels in waves, but later research has modified these theories.

Hooke examined the texture of cork, and found that it was porous, much like a honeycomb, but the pores were not regular. He gave the name 'cells' to these pores, and so introduced the term into the biological sciences. He counted the number of these, and found that there were 60 lengthwise in an eighteenth of an inch, 11,000 in a square inch, and 1,259,712 in a cubic inch, "a thing most incredible, did not our Microscope assure us by ocular demonstration".

Later, on Saturday 10th February, 1678, he noted in his diary, "Examined lobster muscle, found 100 of them in the length of an inch, and the thickness of the threads not more than a 2000th part of an inch, whence there must be 4 million in a round inch." His observations were guided by the principles of matter, motion and number, and although the figures arrived at by his calculations seemed incredible, he was convinced by his observations. The same three factors were to be used by other investigators of the time, such as Dr William Petty in his consideration of social matters.

Hooke investigated Cornish diamants, and considered that the regularity of their crystalline shape could be built up by geometrical arrangement of the spherical bullets then in use from triangles, hexagons, squares etc. Such demonstrations are now used by plastic chemistry and physics sets. He observed that Kettering stone was made up of innumerable small bodies of a globular shape, like the ovary of a herring. This rock is an oolitic limestone made up of a vast number of fossilised marine creatures' eggs.

The last illustration, just to fill up space, was of his observation of the constellation of the Pleiades, and of the moon's surface, showing the crater of Hipparchus. This observation he made in October 1664, using a thirty-foot-long telescope. He pondered about the possible causes of these craters, and carried out an experiment of dropping spherical bullets on to wet pipe clay, and noted that the shape of the depressions was similar to those on the moon. He could not envisage what objects from space could have caused these, although one of his other detailed enquiries was about comets. He then thought the craters might have been formed by explosions from within the moon, and showed that heated soft pipe clay also formed bubbles on the surface like the moon's craters. He did think that if humans were able to look at the earth's surface they might find that it had similar depressions to those on the moon.

One of the first people to buy this book was Samuel Pepys, who sat up until two in the morning reading it.

Hooke gave a copy of the book to Dr Busby, which is now amongst the archives of Westminster School. The Dover Publishing Co. produced an edition in 1956, and the *Impression Anastalique, Culture et Civilisation de Bruxelles* produced a facsimile edition in 1966.

The illustrations show Hooke's ability as a draughtsman, and can be compared with Leonardo da Vinci's anatomical drawings for the skill with which they were drawn, and the care taken to make them accurate. As Hooke stated in the preface, one should look at the objects under the microscope from different angles, and at altered positions of the object, before deciding as to its exact shape. This was particularly so with the head of a fly. The book is a remarkable example of all aspects of the genius of the author.

METEOROLOGY

When Hooke started his diary in 1672, his intention was to have his readings of the barometer and thermometer on the left hand side of the page, and other comments about the day's weather, and the notes about daily events on the right hand side. This scheme was not continued, and when the transcribed diary was published in 1935, the first few months giving these details were omitted.

Details of his wheel barometer were given in *Micrographia* and of his spirit of wine sealed thermometer, and how he graduated it. His discovery of the fact that the beard of a wild oat curled in dry weather, and straightened in humid conditions, gave him the idea of making an hygrometer to indicate humidity and dryness of the atmosphere. He later devised an anemometer.

The combination of these three instruments for making weather records led to the development of the science of meteorology. The interest of The Royal Society in this was shown when they asked Dr Dee and others to make Hooke acquainted with their collections of observations relating to the weather.

A month later, on October 7 1663, Hooke read a paper to the Royal Society on the observables necessary for making a history of the weather, and nearly four years later a detailed statement of his scheme was recorded by Thomas Sprat in his *History of the Royal Society*. These observables were:

(i)   Strength and quarter of the winds
(ii) Degrees of hot and cold in the air
(iii) Degrees of dryness and moisture in the air
(iv) Degrees of pressure of the air
(v)   The face of the sky – cloudy or clear, type of clouds
(vi) Effects produced on other bodies – diseases most rife

(vii) What thunder and lightnings happen
(viii) Any extraordinary tides – comets – exhalations

These details were to be tabulated in eight columns, with a general deduction in the ninth column. Each day's observations were to be on a separate page and the thirty folded pages kept as a record of the month's readings in a form to which easy reference could be made.

The day-to-day readings could be studied and compared, and from such records he hoped that it would be possible to forecast the coming weather. One such instance of him doing so was on 15th July 1677, St Swithin's Day, when the diary records, "St Swithins. a pretty deal of rain. D.H. At Jonathan, foretold fair weather. Barometer 170, Otheometer, Thermometer and magnet noe alterations."

There was no rain noted until 23rd August, when he wrote, "Rainy morn." There had been a drop in the barometer reading from 192 on the 13th July, when the temperature was the same as on the 12th, seven, and the same for the next four days. On the 23rd July, a Westminsterian came to see him about a barometer and thermometer, so he may have convinced others about the fallacy of St Swithins Day, and accepted more reliance on scientific observations and records.

He also made an instrument which he called an otheometer, but there are no details as to its construction, but he noted on 12th July that, "It is now very evident that the Otheometer moves directly opposite to the Barometer."

The only note given in the column for degree of moisture was on 2nd January 1673 when he noted that it was "Dry", this out of more than a hundred entries.

At the time that Hooke made his first wheel barometer, it was considered that the variations in mercury level in Britain were between 28 and 30 inches, and so the first barometers of this type had a dial reading from 0 to 200. From an illustration of a zero reading, this was slightly below 3 o'clock on the dial, but having a circular dial is confusing when Hooke wrote on 6th January 1673, "A stormy morning with some rain. Wind S.W. Thermometer 3. Mercury at noon falling 194. At night 200 " and, on 8th January, "Wind W. Mercury fallen in the morn at 9 to 25 but returning. I suppose it had been lower. Thermometer 2. Much thunder and lightning had been seen this and some preceding nights about 3 pm. W. Stormy N.W. a great storm of snow and haile – the rest of the day fair and sunshine."

Later barometers, such as Tompion made for William III, had a scale reading from 0 to 300. This barometer was seen by Uffenbach in 1710 and was moved to Carlton House Terrace and taken to Buckingham Palace, when that was altered by George IV, and later returned to Hampton Court Palace, where it now is. Another Hooke-Tompion barometer is at Kensington Palace.

The highest reading on the thermometer noted by Hooke was on 16th June 1676, "This was the hottest day of all, Thermometer 13." As described in *Micrographia*, Hooke had made the zero on his thermometer the point where pure water began to freeze. having made a mark at the level of the mercury there, he put the tube in full range of reading. Hooke may have taken normal body temperature as the upper limit, as Fahrenheit was to do later. Hooke also extended the range below zero by immersing the bulb of the thermometer in water to which ice and nitre (potassium nitrate) had been added, a method Fahrenheit was to use to obtain the lowest temperature for the zero of his thermometer.

To record the readings of these various instruments, Hooke made an improved weather clock, based on one made by Sir Christopher Wren. The device had a rain gauge as well, and the readings of each of these meters were recorded by means of pens operated by a string going over a pulley wheel. The weather clock took some time in its development; the first mention in the diary was on 6th August 1678. But he did not demonstrate it to a meeting of The Royal Society until 29th May 1679 when it was set up in Mr Hunt's lodging.

It was designed to keep an account of the quantity and time of all changes that happen in the air, as to its heat and cold, its dryness and moisture, its gravity and levity; as also the time and quantity of rain, snow and hail that fall; all of which it sets down in a paper so as to be legible and clear.

On 19th June Hooke showed it to Sir R. Viner, Sir R. Clayton, and Alderman Morris. Pictures of this clock remain, but there is no example of the clock itself. It may have been put in the repository, and neglected, as Uffenbach found was the case with the general state of the repository when he visited it in 1711. The other instruments invented by Hooke and others may have suffered a similar fate.

At a meeting of The Royal Society on 1st May 1679, Hooke showed his elm hygroscope, which was made of 24 pieces of elm cut across the grain, "the shrinking and swelling of every foot of it made sensible". The hygroscopes made today for use in display cases in museums are usually made of hair or gut, but the principle of their action is the same as the beard of a wild oat.

With respect to section (vi) of the things to be observed and registered: 'Effects produced on other bodies – diseases most rife,' John Locke kept weather records for this purpose, as he thought that there was a relationship between certain illnesses and the weather. He sent oat beards to people who he had asked to keep weather records, and one of them, James Pound, a ship's surgeon who was about to leave on a voyage to China, thanked him for it, and said he had a plan for fitting it in a box for an hygrometer. These weather observations were to be used to make a wide survey of social medicine.

COMETS

There are nearly fifty entries in Hooke's first diary where comets were the subject of observation or conversation.

On 1st March 1665, at the meeting of the Royal Society, Hooke was asked to extract out of his lecture on the late comet and fit it for the press. No details are given in Birch's account, but Pepys who was present at this meeting, wrote that he heard Hooke read a curious lecture about the late comet, proving that this was the very same comet that had appeared in 1618, and that in such a time it would appear again. "Here was a very fine discourse and experiments, but I do lack Philosophy enough to understand them." wrote Pepys.

In his discourse on Comets printed in his *Posthumous Works*, Hooke considered the views of Galileo and Hevelius on comets, and gave the most detail about the comets of 1680, and 1682, which he had observed. He noted Thomas Hill of Canterbury had made observations of that of 1680, and he had received reports from Seignor Montanari of Venice, the latitude of which city he noted as 45 degrees 27 minutes north. Dr Wallis had seen it from Oxford, and Hooke had had reports from Flamstead, who had seen it on 10th and 11th December 1680. Hooke sought the views of ancient men, who had seen the comet of 1618.

Hooke raised a number of queries about comets. "There being a gravitation of all bodies towards the sun, it seems a difficult problem to give a reason for the blaze or tail being nearly opposite to the sun."

"Though if fire be only a dissolution of a body by the nitrous part of the air, it may seem pretty difficult to explain how there should be any fire in those places through which comets have been observed to move. The reflection and refraction of the sun's beams are insufficient to explain the appearance of the comet."

Five difficulties which required an answer were:

1. If the comet is a burning body, or body in dissolution, how comes it can supply so vast a quantity of flame or steaming emanation.
2. If the blaze is from the head, and the bodies moving swiftly through the spaces of the ether, how comes it that the flame or steam does not follow after, or point directly to the way through which the head or body of the comet has passed.
3. If the blaze be an actual flame, and all the bodies we know need nitre or nitrous air to make them, how comes it that there should be such a quantity of nitre in the comet, or in the air, which does not extend more than 50 miles above the earth.
4. Why doesn't the burning take place in every direction from the centre.
5. If the blaze were an actual flame, how could it continue to flame for so long, and to such an extent as to stretch some million of miles, and not be extinct in a very short time.

ASTRONOMY

On 20th June 1666, at the meeting of the Royal Society, Hooke reported that he had observed a new spot in Jupiter different from those which he had formerly observed on that planet, and in another belt. He added that he had seen the satellites of Jupiter with Mr Boyle's sixty-foot glass, as bright as he saw Jupiter himself with the naked eye. He undertook to make observations of the parallax of the earth's orb to seconds; and also to make observations with long telescopes without the use of a tube. He was not successful with the attempt to observe the earth's parallax, and the long telescopes without tubes were difficult to use. He was able to report that he had observed the rotation of Jupiter by means of the spots, and this may have led him to invent his helioscope in an attempt to detect the rotation of the sun by sunspots. In spite of reducing the intensity of the sun's light by triple reflections, he still hurt his eyes in the process.

Some entries from the diary give his other discoveries.

22nd January 1673	"New planet observed about Saturn."
20th March 1674	"Finish the tract of the motion of the earth."
20th April 1674	"At Lord Brounckers. Told him of my hypothesis of the planets, and discoursed of old Bonds."
26th January 1680	"At Jonathon's, Sir Ch. Wren about planetary motion."
4th January 1680	"At Jonathon's, Sir J. Hoskins, Tompion, Trumbol, – perfect Theory of the Heavens."

He had observed the sunrise "very ellipticall", the underside much flatter than the upper on 15th June 1676. On 17th June he noted in his diary that he had discovered a spot in the sun. On 30th July 1676 he discoursed of spots in the sun. There is no mention of this in Birch's *History of the Royal Society*, but there is a note that Mr Colson gave an account of the eclipse of the sun on 1st June 1676, which may have aided observation of it, as sometimes there is a haze in the atmosphere at such an event. Between 31st July and 10th August, Hooke was at Mr Boyle's and showed him the sunspots.

He wrote in his diary on 6th August 1679 that he, "tryd concaves sole reflection, better than Reflex refraction, excellent Helioscope."

So at that date he would still seem to have been trying to improve the instrument. There are twenty-four entries in the diary from 15th June 1674 to 6th August 1679 about helioscopes, but as with many of his entries, not much detail. Like Hooke himself, one has to be a polymath to appreciate the significance of each, both with regard to the development of the instrument and for the particular study for which it was invented.

Hooke thought that telescopic sights were essential for accurate observations, and demonstrated to the Royal Society that the eye, without the aid of these, cannot distinguish any distance in the heavens to less than half a minute, and generally to within a minute. He had criticised Hevelius for using open sights, in his observations of the heavens, and later apologised for his writings in the *Animadversions* against Hevelius and that he had no intention of disparaging either his work or that of Tycho Brahe.

INVESTIGATIONS INTO AIR

The discoveries made by William Harvey, Torricelli and Von Guericke stimulated further experimental work to discover the physical and chemical properties of air, and its biological significance.

Torricelli had shown by inverting a long glass tube filled with mercury, that the fall of mercury apparently left a vacuum in the top of the tube. The Accademia del Cimento carried out the experiment with a tube with an enlarged end in which a bladder was put and noticed that the bladder inflated to a greater volume when the tube was inverted and a vacuum created. Von Guericke developed an exhaust pump using a fire squirt which had been modified, and demonstrated the force of the atmosphere when the air was exhausted from two metal spheres held firmly together. Hooke's improved pneumatic engine was used in many scientific experiments which led to a more detailed knowledge of the air and its properties. Dr Gunter, author of *Early Science at Oxford*, and Curator of the science museum there, considered that no more important work had been done at Oxford. It was the *pièce de resistance* at many meetings of the Royal Society. Thomas Newcomen wrote to Hooke about the possibility of obtaining motive power by exhausting the air from a cylinder fitted with a piston. Hooke wrote lengthy notes for Newcomen, and in the next century the improvements made in steam engines by James Watt led to Britain's dominance in industrial developments.

The first discovery made into the properties of air was the "spring" of the air, the proof of the proportional relationship between pressure and volume known as "Boyle's Law". As Boyle was short-sighted and unwilling to engage in any study where mathematical skill was required, used the term "we", and mentioned employing the assistance of a very ingenious man, it was likely that Hooke made the readings and did the calculations. Boyle also noticed that when the cistern of the mercury tube was put in the receiver of the exhaust pump and the air pumped out that there was always a small amount of mercury which could not be drawn down any further. The height of the mercury remaining was a measure of the effectiveness of the pump, and of the partial vacuum it could produce.

An earlier observation made by Hooke was that the height of the mercury in the Torricellian tube was not related to the moon's motion and phases, and that the tides were not due to pressure on the atmosphere as Descartes had suggested.

After his appointment as Curator of Experiments the first demonstration that Hooke gave was of the breaking of glass bubbles by air pressure, and at the same meeting he mentioned that he had made a condensing engine for compressing air. Many experiments were to be conducted in these two machines to study the effect of reduced or increased air pressure on burning objects and on animals, and on the resistance of air to moving objects.

On 12th December 1662, Hooke gave an account of the rarefaction of air, its expansion relative to the height of the mercury column producing the expansion, and from his results concluded the experiment by giving a calculation of the height of the atmosphere as $25^{23}/_{125}$ miles. Two years later on February 10th 1664 he gave his results of determining the weight of air relative to that of water, which was the usual method of considering weights then. This he did by weighting a bottle of capacity 119 pints and then weighing it filled with water. The weight of the air was $2\frac{1}{8}$ ounces, the water weighed $2017\frac{1}{2}$ ounces, giving a ratio of air to water of $1:949^{7}/_{17}$.

The term "air" was used then to describe any matter in a gaseous state that was present in the atmosphere, or produced from chemical actions, or by the heating of any liquids or solids. After two experiments in which an "air" was produced by the action of vinegar and aqua fortis on oyster shells, the President, the Lord Viscount Brouncker, in reply to the question as to how it was known what was supposed to be air produced by these experiments, stated that, "a body rarefied by heat, and condensed by cold" was air. It was not until Stephen Hales invented the pneumatic trough for collecting gases effectively that the properties of different gases could be studied and their distinctive features noted. Eleven of what we now call elements were then known, but all except mercury were solids at room temperature and it was not until the trough was used that gaseous elements could be identified.

A number of these naturally occurring "airs" were known by miners, and on 3rd December 1663 Dr Power read a paper on "damps" in coal mines and he distinguished three kinds of damps, common damp, suffocating, and fiery damp. The common damp hindered the miners' candles from burning, the suffocating damp stifled and killed men, and the fiery damp was a vapour which came out of the coal, and sometimes ignited of its own accord, sending a sheet of flame along the roof of the pit. Sir Robert Moray gave an account of miners dropping down dead as they stepped into a part of a mine filled with suffocating damp. Mr Vermuyden mentioned a method of reviving men suffocated in coal mines by putting their heads in fresh earth. On 17th December 1662, Hooke reported that he had obtained an air from spirit of wine, and two months later he gave a lecture to the Royal Society on the "History of the Air" in which he detailed the problems of finding out the true nature of air, of what substances or particles it existed, how were these particles formed, and what were their shapes? What was their quantity and extent and how high did the air extend above the surface of the earth? Was air mixed with the blood or humours of the body, and did fishes breathe? Was air intercepted in the cavities of the earth, in the substances of water and other liquors, and in inanimate objects such as plants, wood and stones? These were a few of the questions he asked which he endeavoured to answer, and which also give some idea of the extent of knowledge at that time. At the same meeting Sir Robert Moray described a furnace into which preheated air was blown, "by the means of which furnace, some persons pretend to melt ore into water", and Dr Wilkins mentioned Mr Towgood's sucking pump raising water 42 feet high.

At the meeting on 18th February 1663 another kind of vapour was mentioned by Mr Howard who gave an account of a sulphureous vapour arising out of the ground at Wigan, "which might serve to boil meat", and he also produced a letter from Mr Roger Bradshaigh of Hawksley, near Wigan who mentioned that, on the local estate of Mr Molyneux, a mixture of water and the earth there produced a vapour that would light a candle at half a yard distance.

At a meeting Hooke experimented with a bladder of air to see how long the same air would serve for respiration and found that he could breathe it in five times. This was repeated on 4th February 1663 and Hooke breathed the same air 19 times in 1¼ minutes, and Dr Merret 76 times in three minutes. Dr Merret mentioned a caulker who managed to breathe so well under water, as to stay under for half an hour to repair a sunken ship.

One development from the compressing engine was a wind gun which Hooke demonstrated in November 1663, at a meeting of the Society, when the gun propelled a bullet at a door twenty yards away and made a dent in the door.

The business of diving then took up some of Hooke's time, as well as the effect of compressed air on animals and on burning objects. A mouse was kept in air compressed to one eighth of its volume, and remained alive and well. From this it was inferred that a man could breathe under water at a depth of 100 fathoms "if he can breathe and live with as thick air as a mouse can". On 25th March 1664 a carp bladder was subjected to increased pressure in the condensing engine and it appeared that one part of the bladder was crushed.

At a previous meeting a bird had been killed by the compressing engine, and its heart and lungs were found to be damaged. A week later Hooke gave an account of a bird kept in air compressed to a quarter of its original volume, "somewhat panting". It was then taken out, grew lively, then sickened, but recovered fully. Another experiment reported about the same time was of two birds, one put in a glass of common air and the glass sealed with clay, and the other was put in a glass of common air sealed with cement. Both birds were kept sealed in for eight hours. The bird sealed with clay survived, but the other died. A fortnight earlier he had kept a bird in air compressed to half its volume for ten hours. When the bird was taken out it was lively, but later sickened and died towards evening. In the previous year Hooke had attempted to kill a mouse in condensed air, but this was not successful as the vessel was not airtight.

Experiments were also performed under reduced air pressure. On 1st April 1663, a tench was put in a bowl of water and enclosed in the receiver from which the air was then pumped out. The fish rose to the surface, its eyes swelled, and the operator was instructed to observe the fish and how long it lived. A week later he reported that when, after an hour and a half, he had opened the glass, the fish sank to the bottom immediately and when taken out was found to be dead. The fish was cut open and its bladder was found to have burst. Hooke was instructed to try the experiment with other kinds of fish, and on 20th May 1663 a middle sized carp was put in the receiver, the air exhausted from it, and the fish suffered a similar fate.

Hooke had also put a burning lamp inside the receiver and compressed the air, and found that the lamp burnt longer in compressed than in uncompressed air.

At one meeting after such experiments he was ordered to try some tadpoles in the rarefying engine. He was also asked to observe what Dr Charlton had suggested of their being frogs after the skins were removed.

The Society were also interested in respiration, what its purpose was, and how it was achieved. On 2nd November 1664 Hooke proposed an experiment of opening the thorax of a dog in order to see how long it could be kept alive by pumping air into its lungs, and whether anything could be found concerning the mixture of air with the blood in the lungs. Hooke gave an account of this experiment at the meeting on 9th November 1664. The thorax and the belly of the dog were opened and the greater part of the diaphragm cut away. Air was then pumped into the windpipe, and the heart continued beating all the time air was pumped into the lungs. Even when he made a ligature around the main veins and arteries to the lower part of the body he could detect no change in the animal's pulse. Neither could he detect whether the air joined or mixed with the blood.

Dr Goddard thought that the air might pass through the lungs, but more than ordinary force might be needed, but he thought that if it did in any animal that was unopened, it might hinder the motion of the lungs.

Hooke gave an account of this experiment to Robert Boyle in a letter which he wrote on the following day. "My design was to make some inquiries into the nature of respiration. But, though I made some considerable discovery of the necessity of fresh air for the continuance of animal life, yet I could not make the discovery in this of what I longed for, which was, to see if I could by any means discover a passage of the air out of the lungs into either of the vessels of the heart; and I shall hardly be induced to make any further trials of this kind, because of the torture of the creature: but certainly the enquiry would be very noble, if we could find a way to stupify the creature, as that it might not be sensible, which I fear there is not any opiate will perform."[1]

At the meeting at which Hooke and some other members tried breathing the same air a number of times Hooke listed a number of inquiries to be made about Greenland, and one of these was to consider the anatomy of whales and the contrivance of respiration of other fishes and morses.

[1] New questions relating to respiration following Harvey's discovery of the circulation of the blood can be read in Descartes' Discourse 5.

The nature of air also interested the Society. On 1st July 1663 it was ordered that the council meet on the following Monday to consider experiments to show King Charles on his visit, and Col Long, Dr Wren and Mr Hooke were desired to meet with council in preparation for this. At the same meeting Hooke's experiment to determine whether the bubbles coming out of water as the air was exhausted from the receiver were real air, or only rarefied parts of water was ordered to be repeated at the next meeting. Dr Wilkins also reported that Dr Croune had found an abundance of small insects in the blood of a dog dissected by him, and Hooke was ordered to note that and to make frequent observations with the microscope on the blood of several animals.

On 16th July 1663 Hooke gave an account in writing of an experiment to unite air and water. Six months later, on 4th January 1665 Hooke made an experiment tending to show that air was the universal dissolvent of all sulphureous bodies, and that the dissolution of these was fire; this was due to a nitrous substance inherent and mixed with air. He showed that a burning coal put under a glass vessel soon went out, but as soon as air was admitted, the coal burnt brightly again. On 22nd February of the same year he showed that sulphur dropped on heated nitre in a receiver from which the air had been exhausted burnt vigorously, just as if it were in the open air, and Mr Boyle affirmed that gunpowder would burn inside a receiver from which the air had been withdrawn.

On March 8th 1665 Hooke mentioned several liquors which working on one another would produce air, viz: oil of tartar and vitriol, spirit of wine and turpentine etc. Three weeks later he was asked to give an account of trying to obtain air by the action of aqua fortis (nitric acid) with the powder of oyster shells, and reported at the next meeting that the greater part of the air formed had returned into the liquor.

He was ordered at the next meeting to make the experiment of generating air with bottled ale, supposed to be wholesome to breathe in, which the air generated from aqua fortis and distilled vinegar was not. A week later he tried generating air from bottled ale, corked and tied fast about an ox bladder, but this did not produce any air. It had been ordered at the previous meeting that the air produced from aqua fortis and distilled vinegar be given to a dog to breathe in to see the effects, but no report was given on this request.

Hooke's letter to Boyle from Gresham Street dated 8th July 1665 gave some details of the plague in London, and that Dr Wilkins, Sir William Petty and he had made plans to go to Nonsuch in Surrey as soon as their implements had been sent there. He had prepared a list of experiments, and they intended to continue those on motion through all kinds of mediums. He also asked Boyle to send a list of any he would like them to do for him.

The next letter was from Durdans, the home of Lord Berkley near Epsom, and in that letter Hooke reported an experiment of lowering a set of four candles fixed on a board down a deep well, and at a depth of two hundred and forty feet they went out. They burnt freely until nearly reaching that depth, then their light grew dimmer and suddenly went out, "as if suddenly quenched or extinguished by sinking into a damp". After giving Boyle a list of the many experiments that they had done, Hooke also suggested lowering a man to that depth in the well, "if we can by any means make it safe". Hooke also asked Boyle for a list of any experiments he would like done, and said he was going to the Isle of Wight shortly, and would do these on his return.,

Hooke returned to London in the New Year, and his next letter to Boyle dated 3rd February 1666 also gave details of the experiments in the well; that the air above was very cold, but that at the bottom it seemed exceedingly hot, that the vessels lowered into the well were covered with thick drops of water when they were drawn up, that the lighted candles always went out at the same depth, but relighted if drawn up. He also measured the time taken for wooden and leaden bullets to fall to the bottom, and the time taken for an echo from the bottom of the wall compared with that from a wall above ground an equal distance away. He also mentioned that he was making a collection of natural rarities.

A practical application of supplying air for divers had been an earlier topic of enquiry on 1st May 1661, when the amanuensis had been ordered to provide an engine for diving as soon as possible. Six weeks later he was ordered to use the utmost expedition in making the engine, and on 24th July 1661 it was reported to the Society that the amanuensis had stayed in the engine for eight and twenty minutes under water. A year later Mr Povey reported that Mr Jonas Moore had a diving engine in which a person could stay under water for two or three hours. Hooke spoke of a way of carrying air to the bottom of the sea at any depths, under a bell, and he was asked to bring in a description of his method. On 13th January 1664 Mr Jonas Moore informed the council of the Society of Sir John Lawson's desire that they would appoint a committee to examine Mr Greatorix's diving machine with the object of using it to build the mole at Tangier at a depth of four or five fathoms. At the meeting of 10th February 1664 the President moved that the business of diving might be taken into consideration as a thing that would be very acceptable, if it could be reduced to practice, and Sir Robert Moray, Sir William Petty, Dr Wilkins, Dr Goddard, Mr Oldenberg and Mr Hooke were appointed a committee for that purpose.

A week later Hooke produced his bellows for use under water for taking in air, but also proposed a better way of supplying the diver with air by two cylinders and two pipes by which air could be pumped into one cylinder and out of the other. During the following week the experiment of supplying a bell or hogshead with air by means of two buckets was tried and was successful, whereupon Hooke was ordered to get a model made ready for the next meeting of the bag which was to be about the diver for continual respiration. After the air inside was used up, the diver was to return to the bell to get a fresh supply.

On 10th March Hooke produced the two leaden boxes to be used for providing air through the two pipes, "whilst the diver comes out of the bell or tub and walks up and down working: which air being spent, the diver enters the bell again for a fresh supply".

At this meeting Sir Robert Moray mentioned a person who had done very notable feats in a bell at a depth of 24, 25, and 26 fathoms, such as taking ballast out of a ship and sawing through the planks of a ship in order to get out the guns, but at a depth of 24 fathoms, he used to fall into a swoon and by the compression of the spirits, bleed at the nose and eyes.

On 16th March 1664 Sir Robert Moray reported to the Society, that he had been informed by the French ambassador that there was at that time a diver at Dieppe able to stay under water for one or two hours and take valuable objects out of sunken ships. He had recovered a box containing 40,000 guilders from a wreck. He wore a headpiece wide only about the mouth and a leather suit well stopped and tied about him. Sir Robert suspected that this diver might use a piece of sponge dipped in oil, and held in his mouth. Mr Hill was asked to obtain information about the divers of Toulon, Genoa and Ormuz, and to consult Purchas about them, and the skill by which they stayed in water for some time.

For the Society's experiment Mr Evelyn had been asked to inquire about a diver at Deptford, and was successful in his efforts. Sir Robert Moray also read an account given him by Mr Maule, a diver, of the bell that he used. This bell was about 2½ feet deep and the diver sat on a rope fixed across the bell. The diver never went outside the bell and was able to stay at a depth of 15 or 16 fathoms for half an hour. If the work took a longer time two or three divers would relieve each other in turn. They covered their bodies with train oil, and wore trousers and a close fitting flannel waistcoat. These clothes were dried out over a fire after each man's turn in the bell.

A diver employed by Mr Pepys took part in a later trial after being instructed in the use of the two cylinder apparatus by Mr Hooke. The attempt was unsuccessful, but Hooke left the apparatus with the diver so that he could practise. Hooke had also supplied this diver with deep concave spectacles to aid his vision under water. This experiment was repeated on 1st June 1664, when the diver stayed under for four minutes.

At the meeting on 12th October 1664, the question had been raised by Dr Charlton, Dr Ent and other physicians, as to whether there was any visible passage of air into the heart. Hooke had earlier dissected a viper, and given an account of its poison fangs, and he was instructed to do this further dissection of its lungs. He found that they were one long bladder, extending from the throat to the middle of the belly, and the upper end was a network of veins and arteries. He examined these under the microscope, "and I could plainly perceive all along the sides of these vessels several small chains of bubbles of air, whether they were separated from the air in the lungs to be mixt with the blood, I was not able by this dissection to discover".

There was a lapse in this subject until 23rd May 1667, when Hooke moved that it might be tried by experiment to find out whether it was the supply of fresh air or the motion of the lungs which kept animals alive. He had performed the operation of blowing air into the lungs of a dog in November 1664, and asked to be excused from repeating it, so Dr Balle and Dr King were asked to do so.

Later in the year, in October 1667, Hooke performed the operation with Dr Lower, and they were able to keep the dog alive. Hooke's opinion as to the true purpose of respiration was that it discharged the fumes of the blood. A further conclusion was that the animal could be kept alive without any motion of the lungs, and that these did not contribute to the circulation of the blood.

An experiment proposed for the next meeting was to

bypass the lungs, by making the blood flow directly from the vena arteriosa into the aorta, Dr Lower and Mr Hooke were to take care of the experiment. At the meeting of 23rd April 1668, some criticisms of the experiment were made in a letter sent by Dr Walter Needham to Oldenburg, and during the following discussion on respiration, Mr Daniel Coxe suggested that it was proper to examine and separate the parts of air, in order to know what there may be in it, that may make it so necessary for respiration.

A century was to pass before the isolation of oxygen was achieved.

Earlier that year, in June, following a statement by Robert Boyle that he knew of a man who had stayed under water for three hours, there had been a discussion as to what quality it was that made air fit for respiration. Hooke gave it as his opinion that there was a kind of nitrous quality in the air, which made the refreshment necessary for life, "which being spent or entangled, the air became unfit". It was that discussion which had led to the experiments on dogs mentioned above.

Hooke answered Dr Needham's criticisms, and also proposed an experiment to see whether the blood circulated when the lungs were subsided. He also remarked that it had been observed that the blood, though of a dark, blackish colour, would, when exposed to the air, become presently florid and, the surface being taken off, the subjacent part being exposed again, that would also attain the same floridness; and that, "therefore it might be worth the observing by experiment whether the blood, when from the right ventricle of the heart it passes into the left, coming out of the lungs, it hath not that tinture of floridness before it enters into the great artery; which if it should have, it would be an argument that some mixture with the blood in the lungs might give that floridness".

Hooke was ordered to try in private his suggested experiment of passing the blood directly from the right ventricle of the heart to the left, bypassing the lungs. Hooke reported on 4th November that he had attempted the experiment, which had not succeeded, but he would try another method. Dr Lower, who had been appointed the curator for anatomical experiments, reported that he had caused the blood to circulate in a dog by joining the jugular vein to the jugular artery, and the dog continued well.

A letter produced by Dr Needham written by Mr Templer, dated 30th March 1671, concerning the structure of the lungs, which seemed to him to be a complication of a multitude of bronchia and blood vessels, started a discussion on respiration. Hooke thought that its principal use was the conveying of something essential to life into the blood, and something noisome discharged from it. He wondered whether there were valves in the arteries, and thought these might be discovered by injecting coloured melted beeswax mixed with tallow into the arteries, and this was recommended to the physicians of the Society to investigate.

There were some other strange ideas about the colour of blood. A letter sent from Barbados by Dr Thomas Towne stated among other items of information, that the blood of negroes was as black as their skin.

A year later on 24th January 1678, Hooke produced two experiments concerning the constituents of blood and milk, as had been mentioned to him in a letter from Leeuwenhoek, which parts could be plainly seen when a smear of blood was put on a piece of looking glass plate. From this the blood consisted of two substances, a clear liquid and an infinite number of globular parts of about the same size and moving swiftly in all directions. The globular parts of milk were of different sizes and were white. Leeuwenhoek had estimated the size of the red particles in blood to be 100 diameters the size of a coarse grain of sand, which he had judged to be one-thirteenth of an inch. Hooke had written to Leeuwenhoek suggesting that he correspond with the Royal Society, which he did, and to Hooke directly. In the same letter were Leeuwenhoek's observation on eel worms in vinegar, which was another investigation to which Hooke's attention was directed. Dr Wren had early observed that the blood of flies was white. Wren had also observed that if the head was cut off flies, the body of the fly would run away, and if the tail was cut off, they would live another day, but if they were stabbed in the body they died quickly, as all the muscles were in their breast. Dr King had also reported at the same meeting that he had cut off the head of a cat and taken out its heart, but when some time later its tail was pinched, its body and legs moved strongly.

CLOCKMAKING AND THE LONGITUDE

The discoveries made by the Portuguese and Spanish explorers, and the trade they had developed across the Atlantic and Pacific oceans, set new problems for the navigators in determining their position. According to one of the pilots of Cabral, the discoverer of Brazil, errors of four or five degrees were almost unavoidable. As at the equator this would give an error of the order of three hundred miles, new methods of navigation were essential. A Frisian named Reiner Gemma invented a method which required an accurate measurement of time, but with the unreliability and low degree of accuracy of clocks at that date, 1530, the method was not a practical one. Later in that century Philip of Spain offered a prize of 60,000 ducats to anyone who could discover a way to determine longitude accurately, and the States General of Holland offered a reward of 30,000 florins for the same purpose.

One advance in the possibility of more accurate timekeeping was the discovery by Galileo of the moons of Jupiter, and his realisation that the instant of their eclipse could be used as a way of giving the time, once a table of these eclipses had been made. The difficulties inherent in this method were the problem of holding a telescope steady while on board ship, and also the fact that the moon was not always visible, because of its phases and from cloud cover. Galileo's other contribution to the improved accuracy of timekeeping was his idea of using a pendulum as a regulator for the mechanism of clocks of which a drawing of his survives, but it was the Dutch scientist, Christian Huyghens who first made a pendulum clock in 1657. A pendulum clock was later used by Jean Picart to make a more accurate map of France, but the problem of using a clock on board ship was still subject to errors due to the motion of the ship.

Hooke's interest in this was twofold. Firstly, improving the accuracy and reliability of clocks generally, and secondly their use for determining longitude. Before 1660 he had made two improvements to the design of clocks. The first was the application of a spiral spring to the arbor of the balance wheel of a clock, to use the spring's natural oscillation as the regulator of the mechanism. The second improvement was the design of the anchor escapement, which was better than the verge or crown escapement then in general use, as can be seen in Huyghen's diagram of a pendulum clock. Hooke had been advised to patent the clock incorporating his two inventions, but as the proposed patent had a clause which would give the financial rewards to anyone making further improvements, Hooke declined to patent. He demonstrated the spring-regulated watch to Lord Brouncker and others, and was again advised to patent the invention, but again declined.

From an account in *The Philosophical Experiments and Observations of Robert Hooke* compiled by William Derham, some trials were made in 1662 of using a pendulum clock at sea. This was done by Lord Kincardine who had two pendulum clocks fitted below decks, and at right angles to each other. The cases were made heavy with lead, and suspended by ball and socket joints. The clocks had short half-second pendulums. The clocks worked without stopping during the voyage, but did vary with each other. Sir R. Holmes also took two similar clocks on sea trials, and found that he was able to judge his position more accurately than masters using traditional methods. Hooke's experience on the first voyage may have inspired him to invent a more reliable clock, as a paper written by him in 1663 describing a marine clock was discovered in the library of Trinity College, Cambridge by Rupert Hall three hundred years later. A member of the staff of the Science Museum, London, made a clock from this description, and displayed it at a conference on Hooke held at the Royal Society in July 1987.

On 22nd January 1674, Hooke was asked to employ a craftsman to make a quadrant which he had designed, at a cost of not more than £10. He may not have been satisfied with the work of Tom Shortgrave, the instrument maker employed by the Royal Society, and employed Thomas Tompion.

By 5th July 1674 the quadrant was finished and brought to Gresham College where it was seen by Sir Jonas More, and on the following day Hooke tested it and found it satisfactory. Tompion was paid £5 towards its cost on 22nd August, and another 30 shillings on 15th October.

This quadrant is shown as Figure 1 between pages 428–429 of *Early Science at Oxford* by Dr Gunter, and also in *Thomas Tompion and his Work*, by R. W. Symonds, Batsford, page 18.

On 8th March 1675 Hooke showed Tompion his way of fixing double springs to the inside of the balance wheel of a clock. On 14th April 1675 he wrote: "he tryd double perpendicular spring, did well".

The entry for 12th May reads, "Lodowick here about Longitude. Afirmed £3000 premium and £600 more from the States (General of Holland)."

He presented the King with a watch on 17th May 1675, and on the following day he met the King in the park, who affirmed that the watch was very good. He received the watch back and proposed longitude by time. On 25th May he recorded in his diary that he received the double-pendulum sea clock, had a box made for it by Coffin, and hung it up by strings. He showed it to Tompion on the Sunday following, and to the King and Sir Christopher Wren on the Monday. Both liked it, as did Sir Jonas More.

It would seem from these entries that Hooke made a clock such as he had described in his paper of 1663, and with the improvements noted above. He had hoped to benefit from this invention, but it was to be another hundred years before John Harrison made a clock sufficiently reliable to be used at sea for determining longitude.

Hooke's first discussion with Tompion about watches took place on May 2nd 1674

"To Thomkin in Water Lane. Much discourse with him about watches. Told him the way of making an engine for finishing wheels and how to make a dividing plate; about the forme of an arch; another way of Teeth work; about pocket watches and many other things."

The footnote to this page states: "Thomas Tompion, the Father of English watchmaking."

Tompion no doubt owed some of his success to Hooke for revealing his inventions, but Hooke was fortunate also in having such a master craftsman to materialize them.

20th October 1674 "At Anis and Tompions. Show him way for pallets not to stop work."
13th December 1674 "Gave Tompion a description of Aequating time for Sir J. More's clock."
3rd January 1675 "At Sir J. More's. Found his clock faster than the Sun between 3 and 4 minutes."

Later in 1675, Tompion devised a deadbeat escapement which is illustrated in the article by Derek Howse, *The Tompion Clocks at Greenwich, and the Dead-Beat Escapement* and published in *Antiquarian Horology*, December 1970.

Other entries in the diary show other ideas given to Tompion by Hooke.

8th March 1675 "At Garaways. Tompion. I showed my way of fixing Double Springs to the inside of the Ballance wheel thus."
27th March 1675 "Showed Tompion the double pendulum."
13th April 1675 "At Tompions. Tryd perpendicular spiral spring."
14th April 1675 "Tryd double perpendicular spiral spring

did well."
25th May 1675 "I received the Double pendulum Sea clock and had Coffin make box for it. I hung it up by strings."
13th June 1675 "Tompions. One day this week I revived my old contrivance for pocket watches by cutting the Balance in two and inserting the two halfs joyned by two side pieces. Not alterable by any, they fitted the Deans watch well – Tompions – Resolved to proceed."
28th June 1675 "At Tompions. He mended watch with 3 circulating balances."
30th August 1675 "Contrived to move the double balance without teeth. All springs at liberty bending equall Spaces by Equall increases in weight."[Hooke's law].
10th October 1675 "Invented the fly to move circularly for a sea watch."
12th October 1675 "Invented the best way for a circular fly."
18th October 1675 "Completed the theory of fly watch."
15th November 1675 "Meditated about motion of planets of circular pendule."
17th December 1675 "Fell out with Tompion."
24th December 1675 "Tompion a slug"
31st December 1675 "At Garaways. Thompion. A clownish, churlish dog. I have limited him to three days and will never come near him more."
5th January 1676 "Tompion a rascall."
6th January 1676 "Calld on Tompion."
22nd January 1676 "Calld on Tompion."

From these last two entries it is clear that Hooke's annoyance soon subsided as it did it many other expressions of anger in the diary.

He also continued to discuss new ideas with Tompion.

28th March 1676 "Directed Tompion about sound wheels. The number of teeth."
2nd June 1676 "With Godfrey and Tompion at play. Met Oliver. Damned dogs. *Vindica me Deus*. People almost pointed."

This was Shadwell's play *The Virtuoso*, which R. S. Westfall, who wrote the biography of Hooke for *The Dictionary of Scientists*, described as "a wretched physico-libidinous farce".

14th September 1676 "At Tompions, where watch with endless screw."

15th October 1676 "Tompions about Dr Busby's watch. I taught him the way of the single pallet for watches. Discoursed about laying ballances one over tother."

10th November 1676 "At Tompions, told him my way of new striking clock to tell at any time hour and minute by sound."

21st June 1677 "A stranger showed me a watch with a spring a hair. Tompion here. Instructed him about the King's striking clock, about bells and about the striking by the help of a spring instead of a pendulum as also the ground and the use of a swash teeth."

29th June 1677 "Contrived motion of the pendule without noyse by universal joynt."

Today the universal joint is used in the driving shaft of a car for linking the engine with the rear wheels in rear wheel driven cars, and also in the supporting blocks for an oil rig, where the joint is as large as a house.

7th July 1677 "I showed Tompion a second way by cranks to make a pendulum goe without noyse."

12th August 1677 "Scarborough here. Resolved on watch with two springs bende about cylinders and a 3rd spring regulator moved by strings."

Hooke also records discussions with Bennett, another clockmaker, who did work for him, and Bell, from whom on 20th March 1673 he "borrowed Bell's engine for cutting wheels. He told me of one he had to perfect them but would not show it".

Hooke made a wheel-cutting machine, and such a device

improved both the speed and accuracy with which the train of wheels used in a clock could be made. R. W. Symonds in his biography of Thomas Tompion states that Tompion made 6,000 watches and 550 clocks. To achieve this output he organised his workshop in a way previously unknown in English handicrafts by using the principle of division of labour advocated by Sir William Petty.

"In the making of a Watch, if one Man shall make the Wheels, another the Spring, another shall engrave the Dial-Plate, and another shall make the Cases, then the Watch will be better and cheaper, than if the Whole Work be put upon any one Man."

Thomas Tompion had been trained as a blacksmith by his father, who had his smithy at Biggleswade, and the two quotations from contemporary journals record this.

"The great Tompion had never made Watches, had he not first made hobnails."– *The Weekly Journal* or *British Gazeteer*, 17th January 1719, and the second, "When you next set Youre Watch", wrote Matthew Prior (1664–1721) "remember that Tompion was a farrier, and began his great Knowledge of the Equation of Time by regulating the wheels of a common jack to roast meat".

EXPERIMENTS IN BRICK MAKING

At the meeting in October 1666, it was moved that the materials for building, and the several sorts of earth for making brick and tile, might now be considered by the Society, who were desired to think about it against the next meeting. It was mentioned that there was many a good terrace in England, especially in Derbyshire. On 31st October, mention being again made of considering the several sorts of clay fit for making bricks, Sir Paul Neile affirmed that there was a certain clay in England which made as good sounding bricks as any of those called klinkers in Holland. The Earl of Kincardine remarked that the klinkers in Holland differed from the other bricks chiefly in the manner of burning, those that lie near the fire making the most lasting bricks, the remoter the softer. Another member mentioned that Mr Wylde had a way of mixing several sorts of earths together, to make hard and lasting bricks. Mr Hooke took notice that those earths that will vitrify make the most lasting bricks. It was ordered that Mr Hooke should make trials of several earths by burning them in a wind furnace, to see which kind would make the best bricks.

On 28th March 1667 Mr Hooke proposed an expeditious way of making bricks the consideration of which was referred to the next meeting. A week later he was ordered to produce his method of making bricks with less charge and more speed than had hitherto been used. Dr Wren mentioned that he had seen a digging machine contrived by one Mr Bayley, which would perform twice as much as the same number of men in the usual way. On 11th April 1667 Mr Hooke was put in mind to bring in his model of a machine for making bricks expeditiously. On 9th May the machine was produced, and tried with some clay but, that being too stiff, the trial did not succeed. The members discussed the method, but considered that vast areas of land would be required to lay out the bricks so made, so thought it best to lay the idea aside.

Three years later, on 1st July 1669, a letter from Mr Huyghens was read in which he gave a description of a burning concave mirror of thirty four inches in diameter, which could vitrify a brick in a minute.

MEASURING A DEGREE AND THE CIRCUMFERENCE OF THE EARTH

At the meeting on 23rd January 1667 Hooke affirmed that the circumference of the earth might be measured to seconds by a sixty foot glass put perpendicularly at a place where the distance may be conveniently measured, such a one as may be smooth and a mile long, lying north and south, or, at least, north east and south west. He was ordered to make this experiment as soon as a convenient place could be found.

In April he proposed a way of measuring the circumference of the earth with a twelve foot glass and three stakes, to be practised in St James's Park on a calm day. On 2nd May it was ordered that the telescope and some stakes should be ready for the following Monday morning in St James's Park by the canal for carrying out this experiment. The experiment was not done on that day nor on several subsequent occasions.

At the meeting of 3rd October Hooke was ordered to prepare to open the thorax of a dog at the next meeting, and provide a dog for the purpose. He was also asked to prepare his rarefying machine capable of holding a man, and also, especially, to endeavour to make the experiment for measuring the compass of the earth, moved so long ago and pressed for so often.

At the next meeting Hooke was asked to peruse some papers about sugar, and consider what use they might be for making a history of sugar. He was also asked to bring in a history of soap making and hat making. The attention of the Society was then directed to blood transfusion and respiration.

At the meeting before the summer vacation in 1668

Hooke was again ordered to make the experiment in the park to determine the circumference of the earth, and also to observe the parallax of the earth's orb. On 22nd October, the first meeting after the vacation, the president stated that he understood that Mr Hooke had erected a tube to try whether he could observe to a second the passing of any fixed stars over the zenith, and thence find a parallax of the earth's orb, in order to determine the earth's motion. Mr Hooke proposed that the experiments might be done in order to state the laws of motion, but the president said that the members ought to consider whether this was necessary, as Mr Huyghens and Dr Christopher Wren had already taken great pains to examine this subject, and were thought to have a theory to explicate all the phenomena of motion. Mr Oldenburg was ordered to write to Huyghens and Wren and acquaint them with what had been said at the meeting, and to ask whether they intended to publish their speculations and trials of motion, and assure them that their findings would be registered as their own.

A year later, on 21st October 1669, the Lord Bishop of Chester acquainted the Society that his Majesty had expressed a desire of having the measure of a degree upon the earth measured, and expected the assistance of the Society in it. It was then decided to form a committee of the bishops of Salisbury, Chester, Sir Robert Moray, Sir Paul Neile, Dr Wallis, Dr Christopher Wren and Mr Hooke, or any three of them, and meet at the president's house in Covent Garden on the following Monday evening at five, and make a report to the Society. Mr Hooke was asked to read what Ricciola had written on the subject.

This committee being asked at the next meeting for their opinion as to the best way to proceed, Hooke gave it as his opinion, that one of the most exact ways of making this measurement was by taking accurate observations using a perpendicular telescope by which angles accurate to one second could be measured. At the next meeting Hooke explained his way of dividing a degree into very many minute parts, which he thought easier than by a screw or a sliding ruler. The method consisted of proportioning a short line to a long line. The method being approved, it was thought to be suitable for use in making the experiment which his Majesty had requested, and Hooke was ordered to make the apparatus with all possible speed. Hooke produced this at the meeting of 25th November, and it was so made that any unevenness of the ground would not affect its accuracy.

Hooke showed this instrument on 2nd December, and how the two inches on the scale could be divided into thirty parts, the two inches representing a degree, so that a second could be measured. The instrument was used in making two observations, one northward of the zenith, and its angle with that, and then a measurement southward. Hooke excused himself for failing to have the experiments appointed for the next meeting, as he had some avocations of a public nature which had taken up his time. None of the appointed experiments were related to the measuring of a degree. He was asked to do this at the meeting of 9th December, together with completing his clock which would go for fourteen months, continue with his method of staining, and his new way of printing.

Before the next mention of the degree Hooke produced his new method of staining cloth, his suggestion for a universal measure using a ball of mercury dropped on a dry surface until the horizontal diameter was double its perpendicular height. Another way was of counting the number of drops of distilled water to measure an inch. In the first case the longer of the two measures was to be the standard inch. Hooke also gave his opinion on Dr Wren's machine for grinding hyperbolic lenses. He also produced his mechanical muscle. This operated by applying heat to a body filled with air to produce dilation, and cooling to make it contract. He suggested that if this made the muscle move slowly, the action could be speeded up by using springs. Dr Goddard also carried out experiments using an arm made of tin.

There was an interval of another six months before the

next stage in the attempt. On 9th June 1670 Hooke was ordered to make ready for measuring the degree at the Bedford River, which was a straight stretch of water extending for twenty miles, formerly surveyed with exactness by Mr Jonas More. He had been in charge of the drainage of this area, and it had been the making of his reputation as an engineer and practical mathematician.

Hooke promised that he would do this at the first frost and clear weather when reminded of the experiment on 27th October.

Six months later he engaged to do this within a month. He was also ordered at the same meeting to continue his efforts to observe the parallax of the earth's orb, to finish the burning concave mirror to focus all the sun's rays at one point, to prepare the glass bell and powder to demonstrate what we now call Chladni's figures, but which were, in fact, discovered a hundred and twenty years before Chladni re-discovered them. This was an attempt to show the internal motion of bodies. The final task was to fix a receiver to the pneumatic pump so that a person's arm could be put inside, and the effect noticed.

The next reminder was on 11th January 1672 when he was exhorted to pursue and finish his way of measuring a degree, which he promised to do, hoping to bring it to a greater exactness and nearness. At the same meeting an observation made by Cassini at Paris of a new planet near Saturn was reported, the details of which were given to Hooke for him to make like observations.

At that meeting it was also reported that M. Picart had measured a degree and that he calculated it to be 57060 Parisian *toises* or fathoms, or seventy three English miles. According to the description given in Lloyd Brown's book *The Story of Maps*, Picard used a much more complicated method involving surveying thirteen great triangles between two points. The method suggested by Hooke was based on the geometry of a chord, either the rule-of-thumb application of that used by bricklayers in building an arch: "Square half the span; divide by the rise; add the rise; then divide by two. This gives the radius of the arch."

The geometrical study of the chord of a circle gives the same result. Hooke's first idea of using three sticks and a telescope seems to imply he was going to use this method, and although the distance suggested at first, the canal in St James's Park, would not have been possible, the later place mentioned, the Bedford Level, could have given a reasonable result.

The only evidence that this was eventually done was recorded in the autobiography of Alfred Russell Wallace, who was trained as a surveyor, and who did this experiment as the result of a bet made with a flat earthian to prove that the earth was round. The entry does not give his measurements, which could have been used for measuring a degree.

There was a lapse in interest in this project until 7th June 1682 when, following a discussion on the comparative measurements used in different countries, Hooke mentioned the care taken by M. Picart in comparing the measures of length of other countries with the Paris foot in his book concerning the measure of a degree on the earth. The president then inquired of Hooke why the measure of a degree had not been done, as had been requested. Hooke replied that he would willingly do this if the Society would defray the cost. He also pointed out that the result obtained by Picart differed very little from that obtained by Mr Norwood from measurements made between London and York, and if a more accurate result was desired the Society should first have in their possession an accurate measure of the standard foot of London, and that for all experiments they should have the true weights and measures of England. Hooke was asked to procure those for the Society. The president was of the opinion that the best standard for this purpose would be a certain part of the length of a degree upon the earth, if accurate trials could be made in several latitudes, and they should be found the same, provided the earth was globular and not oval.

The next mention of the measurement of a degree was on 23rd June 1686, when Halley was offered £50 or fifty copies of the *History of Fishes* to measure a degree to the satisfaction of Sir Christopher Wren, the President, and Sir John Hoskyns.

Mr William Molyneux of Dublin, in a letter to Halley on 4th August 1686, wrote that he found the performance of Mr Hooke's level fully satisfactory, and also that he was glad that the Royal Society had given Halley the task of measuring the degree. He mentioned Norwood's attempt and that of Picart, Fernelius and Snellius, and that the last named was likely to be the most accurate as the country was so flat.

GEOLOGY

Hooke's lectures on earthquakes which cover one hundred and twenty pages in his *Posthumous Lectures* show him to have been among the first of the geologists. In his diary he had noted many instances of earthquakes, as when on 9th October he met Mr Ludowick at Garaways, who gave Hooke an account from India of an earthquake. On 12th January 1676 he had a letter from Cole about an earthquake in the west. At the meeting of the Royal Society on 31st October 1678 he read a paper to them about the Palma earthquake. In 1693 there was news of one in Malta. There was a discourse with Beaumont about coral and petrifications. These led to his interest in the origin of the earth, and the changes which must have occurred since the Creation.

He had noted fossil remains and had several sent to him by Fellows of the Royal Society. He had observed a layer of sea sand above water level in the Isle of Wight, in which were embedded oyster shells, limpets and periwinkles, and this band of rock extended to a height of sixty feet. Hooke also reported an observation by Dr Peter Ball when passing through a narrow way in the Alps at a height of several thousand feet above sea level where he saw similar kinds of fossils. Hooke had seen marine fossils in a brickfield at St James's Square many feet below ground level, and he knew of similar fossils found ninety feet deep in a well sunk in Amsterdam. Dr Thomas Browne of Norwich had sent to the Royal Society a large bone exposed after a cliff fall at Winterton, parts of a whale discovered at Wells-by-the-Sea, and bones of a hippopotamus found at Chatham. Another of Hooke's observations was of a cliff along the south coast, four miles long, in which some of the rock layers were horizontal, some sloping and some vertical. These finds posed many problems. What changes had occurred to the earth's surface by which marine shells had been lifted to great heights in mountain ranges such as the Alps, Pyrenees, Appenines and Taurus Mountains, and many miles from the sea, and from each other?

The observed changes of which records existed from historical times were earthquakes, volcanic eruptions and floods. There were records of the first two from places as far apart as Java, Peru, Italy, Turkey and Iceland. The account of Noah's flood, and the origin of the earth, as related in Genesis was the subject of several of Hooke's lectures, and discussions in coffee houses.

The story in Plato's *Timaeus and Critias* of the great eruption causing the 'Lost Atlantis' and added to in Ovid's *Metamorphosis*, when, "there was an earthquake and floods of extraordinary violence, and in a single dreadful day and night all your Athenian fighting men were swallowed up by the earth, and the island of Atlantis similarly swallowed up by the sea and vanished. This was why navigation was hindered in the Atlantic by mud just below the surface, the remains of a sunken island."

This last may have been a fable fostered by the Phoenicians, who had ventured into the Atlantic, and wanted to deter others from doing so, as in a similar way they tried to hide from other traders the source of their tin from Cornwall.

That there was an immense eruption at Thyra, Santorini, in 1350 B.C. which destroyed the Minoan civilisation, was an established fact, and there were further eruptions in 196 B.C., 726 A.D., and in 1650 A.D. which were known by Hooke.

The other great natural catastrophes were floods. Hooke knew of the deposition of soil by rivers such as the Nile, but in considering Noah's flood, which must have occurred sixteen and seventeen hundred years after the Creation, according to the first chapter of Genesis, he changed his mind during the course of these lectures. He had thought that this flood, which covered the highest mountains, was the probable cause of the dispersal of sand and shells over the entire earth, but in a later lecture he changed his mind as he considered that two hundred days of Noah's flood would not have been a long enough time for so many shells to be produced, nor for the thickness and quantity of rock to be deposited.

As to the age of the earth, he was doubtful of that as estimated by the Egyptians, nine thousand years before Solon visited Egypt and at the other extreme, the Chinese estimate of 88,040,000 years. Had he speculated on the time taken for the rock layers that he had observed on the South Coast to be deposited, it might have caused him to alter his ideas as to the time scale. It was to be another hundred years before James Hutton's observations and theories altered scientists' views as to the age of the earth.

There are many references in the diary about fossils and petrifications, and he discussed the creation of the earth, as it is described in Genesis Ch. 1 in several of his lectures to the Royal Society, the first of which was given in 1668, and the last in 1697, so it was a subject to which he gave a great deal of thought. He did not change his views as a devout Christian, which he summed up at the conclusion of his lecture on Comets and Gravity.

"I could go through the whole history delivered in the first book of Genesis, but that I only arrived at present to show, that nothing of what I have hitherto supposed, does in any way disagree with Holy Writ, but that it is perfectly consonant with it, as well as it is to Reason and the Nature of things themselves."

He noted that some of the fossils he saw were petrified, whereas others were impressions in the surrounding rock. He thought that the petrification might be due to heat, cold or the pressure of the rocks. It might also be caused by petrifying liquids such as he had seen in Wookey Hole Cave, and which he had been informed also were present in caves in Derbyshire, such as Poole's Cavern or where the formation of stalactites and stalagmites could be seen, but where as yet no idea as to the length of time required for these to form was known. There were petrifying waters at

Knaresborough in Yorkshire, and a field at Apsley in Bedfordshire where a crystalline crust was known to form on objects put in it. Hooke also thought that there were petrifying liquids in chalk to produce the flints found in that particular rock. He did not know of any chemical method by which these petrifying processes could be repeated, but he did know that the vitriols, nitres and alums which could be made by experiments all had a distinctive crystalline pattern, and that if the outer layers were removed from such crystals, the same pattern could be seen in the underlying layer and repeated throughout the crystal. Sir William Petty thought that such distinct structures could be used to identify substances.

Hooke thought that the personal inspection by students of rocks and fossils was a necessary part of their education and that verbal descriptions were not sufficiently thorough; nor did they produce an active response from the students, as was more likely to happen when the fossil was handled and its detail studied. This was, and still is, a sound reason for the study of geology, since it also brings together several scientific disciplines as well as bringing in to play the natural desire to collect things, and fossils have a great interest, as is shown by the present one in dinosaurs.

As curator to the Royal Society, for forty years, Hooke was responsible for storing and labelling the many fossils and rocks brought to him by the Fellows. This collection may have been lost when the Society moved to Crane Court, or dispersed after Hooke's death. Such a collection would have maintained an interest in the subject, as Hooke had intended, whereas there was a lapse of eighty years before Hutton's founding of the science of geology.

Hooke's lectures on earthquakes still give modern scientists food for thought, for example:

Yushi Ito *Hooke's Cyclic Theory of the Earth in the Context of the 17th Century England,*
E.T. Tozer *Discovery of an Ammonoid Specimen Described by Robert Hooke*

D.R. Oldroyd *Robert Hooke's Methodology of Science as Exemplified in his Discourses on Earthquakes.*

The illustrations with this chapter show the skill and attention that he gave in examining each specimen thoroughly, and in drawing them.

THE WAYWISER

The term "waywiser"was derived from the German *weg weisser* or journey measurer, and various methods were used to try to measure the length of journeys. One way was by using the oscillation of a pendulum, as is done in present day pedometers. A way of measuring distances between towns could be done as Mr Norwood did in 1635 when he measured the distance between London and York using a chain measure 99 feet long, as Hooke described in his lecture on navigation. This must have been a laborious task, and was done as part of Norwood's measurements for determining the length of a degree.

The entries in the *Transactions of the Royal Society* show the stages in the development of Hooke's ideas and instruments for measuring distances: "On June 8th 1671, Mr Hooke produced an instrument for surveying, to be applied to chariots, whereby that line or angle shall be made by a chariot thus fitted, shall be described on paper."

He was ordered to get a chariot made and to apply this instrument to it, against the next meeting. He mentioned that he had a way to show the several quarters of the world in a travelling chariot, so that wherever a person goes, he shall have a hand always standing north and south. He desired to produce it before the Society.

On 15th June 1671, the experiment appointed at the last meeting of testing the new surveying instrument to the wheels was made with good success, and it was thought that if the whole apparatus necessary for it to be accurate, it would answer to the design. Hooke was desired to bring in a description in writing.

"November 23rd 1671. Mr Hooke produced an instrument contrived by himself to show the point of the compass in which a person travels. He was desired to bring in a description in writing; as also to compound this with that instrument which he had produced before whereby the way of a traveller may be traced upon a piece of paper, that so, by one and the same instrument, a traveller may make a map of the country through which he passes."

The invention of the waywiser, the forerunner of the surveyor's wheel, was a paramount achievement which resulted in the production of accurate maps in the seventeenth Century. This point was made by Sir Clifford Darby, a former Professor of the History of Geography at Cambridge University, who according to his obituary in the *Independent*, 28th April 1992, had transformed the study of historical geography. In his book, *An Historical Geography of England before 1880*, Sir Clifford stated that the invention of the waywiser had resulted in the most accurate maps of England and Wales at that time.

John Ogilby's *Britannia* consists of 100 plates of the principal roads of England and Wales between the most important towns.

In the preface to the book Ogilby stated: "we have been much facilitated in this Great Work by the Wheel Dimensurator, which for ease of and Accurateness, as being manageable by a single person, Measuring even the smallest Deviation on the Way, and finishing a Revolution but once in 10 Miles, We readily acknowledge and even in the Wheels themselves, commend rather the Foot-wheel here mentioned, of half Pole Circumference, with the Way-wisers as they are now regulated, than any such like Coach or Chariot Mensurator whatso ever."

In the preface of Aubrey's *Brief Lives*, the editor Oliver Lawson Dick wrote that a new edition of Camden's *Britannia* was foreshadowed, and a set of queries relating to it was printed and considered at several meetings by Christopher Wren, John Hoskyns, R. Hooke, J. Ogilby, John Aubrey and Gregory King. This was after Aubrey had returned from his perambulations in Surrey in 1673, which he had undertaken for Ogilby's proposed map of England and Wales.

There are 120 entries in Hooke's diary relating to Ogilby,

and the entry for 15th January 1674 reads, "With Ogilby to Spanish Coffee House and Dr Croon. Trd dancing shoes (received the previous day) At home. Running gallery till I sweat. Experiment exceeded beyond hope. Contrived pacing saddle with waywiser."

Many of the other entries relate to discussions about the production of maps. Ogilby paid tribute to this machine for making the work so much easier, but did not name its inventor, but one could reasonably conclude from the entries, and Hooke's mechanical inventiveness and skill, that the invention was his.

The total distance measured along the 100 journeys was 8,418 miles and 3 furlongs, and from the illustrations in the book it is likely that Ogilby trained and employed several teams to do this work, in view of the mileage. The illustrations also show a man pushing the waywiser, and another man with a magnetic compass.

There is an example of such a waywiser in the Science Museum, London, and another in the Museum of Local History at St Albans. This one can be seen close to, and the scale clearly seen and examined The circular scale reads up to 10 miles, and is subdivided into miles and furlongs.

In his lectures on Navigation, Hooke suggested making a waywiser capable of measuring distances at sea, a device which could be dragged through the sea, so making blades rotate, and by a system of gear wheels, register on a dial.

SURVEYOR AND ARCHITECT

The Great Fire of London, 2nd–6th September 1666, opened up a new field for Hooke's genius to display itself. The Fire destroyed 13,200 houses, The Royal Exchange, The Guildhall, The Custom House, The Post Office, St Paul's, 87 City churches, and the halls of 44 City Livery Companies. The household effects and goods were valued at £3.5 million. The total cost was estimated at £10 million at that time; four-fifths of the area of the City had been destroyed, stretching from the Tower in the east to the Temple in the west. The City was assessed at approximately one-fourteenth of the direct taxes of the kingdom.

To continue the administration of the City, the authorities took over part of Gresham College; the General Post Office used premises in Bishopsgate, and the Custom House some in Mark Lane.

As the most important centre of commerce and industry in the kingdom it was essential that rebuilding should start as soon as possible. It was also important to take the opportunity to widen the streets, re-site the markets and conduits which impeded traffic, widen the City gates and build houses with materials which were less likely to assist the spread of fire.

A detailed history of the damage and the problems facing the City authorities is related in Professor Reddaway's book, *The Rebuilding of London after the Great Fire*. The City's annual average expenditure was £24,901 18s 6½d and its revenue £13,757 7s 1½d. These figures are given because it is often regretted that the City was not built according to one of the plans submitted to it. Wren submitted a plan, which has been preserved, and so did John Evelyn, which also exists. He presented his to Charles II on 13th September and this was well received by the King. Evelyn had spent some time in Holland during the Civil

Wars, and had been impressed by the brick built houses, and well laid out streets. He had also been present at discussions with the King before the Fire to consider ways of dealing with the nuisance of smoke, and the widening of streets. Hooke had prepared a plan, which he showed to the Royal Society on 19th September, and which was also seen and approved by Sir John Lawrence, a former Lord Mayor, and other aldermen, and was preferred to that of Peter Mills, the City Surveyor.

Richard Waller, in his life of Hooke, stated that he had not seen the plan, but understood that it involved having all the streets from Leadenhall Corner to Newgate in straight lines, with the other streets at right angles, and all the churches, public buildings and market places in suitably convenient places. As with the plans of Wren and Evelyn, "this would have added much to the beauty and symmetry of the whole".

Waller also wrote that how Hooke's plan came not to be accepted he did not know, but it is probable this might contribute not a little to his being taken notice of by the magistrates of the city, and soon after made Surveyor.

A letter from Oldenburg to Boyle dated 18th September 1666, stated that, having seen Wren's model for rebuilding the City, "I then told the doctor (Wren), that if I had had an opportunity to speak to him sooner, I should have suggested to him that such a model contrived by him, and reviewed and approved by the Royal Society or the Committee thereof, before it had come to the view of his Majesty, would have given the Society a name and made it popular, and availed not a little to silence those who continually ask, What have they done?"

There were differences of opinion in Parliament as to which plan to use and what materials, so the problem was left to the Privy Council and the Common Council of the City. The King appointed Christopher Wren, Hugh May and Roger Pratt as his Commissioners to work with such surveyors as the City might appoint. The City appointed Robert Hooke, Edward Jerman, and Peter Mills. The commissioners and surveyors were to meet to discuss alterations to streets, buildings, markets and materials. The rebuilding committee met three times a week, and by 30th November 1666, had a draft ready for consideration. During this time the clearing of the debris was already proceeding, having been commenced on 20th November, after the Common Council, together with Wren and Hooke, had agreed about the measuring of the sites, streets etc. By the first week in December all of the rubbish had been cleared away at a cost of £103 15s 4d.

By the Act for the rebuilding which was given the Royal Assent on 8th February 1667, the general layout of the City was to remain, but the streets were to be widened, the houses were to be of three standard types, they were to be built in brick or stone, and with tile or slate roofs. There were to be no overhanging jetties and no wood was to be used on the exteriors of the house. The height of houses, thickness of the walls, number of storeys and depths of cellars were all defined.

There was to be no "Jerry Building", fines were to be imposed for infringements, and Peter Mills and Robert Hooke had the task of checking for any infringements. It was left to individual householders to rebuild their houses, but no one could commence building until a surveyor had viewed his ground and seen that the party walls and piers were correctly set out. The householder was charged a fee of 6s 8d for this. No building could be marked out until the new street alignments had been staked out and this was not completed until April 1667.

To raise money for the City's expenses towards the rebuilding a tax of 1 shilling a chauldron (34 cwt) of coal was imposed on all coal landed at London. The delivery of this was hindered when the Dutch made an attack on the fleet at anchor in the Medway, and then blockaded the mouth of the Thames. This tax proved to be quite inadequate.

There are more references to Wren in Hooke's diary than to any other person. Many of these meetings were to discuss the rebuilding of churches and St Paul's cathedral. There are also more than a thousand entries relating to viewing sites to make a report, issue a certificate, measure a site or set out the ground for rebuilding, note a nuisance or report any irregularity.

The fees that he received for these surveying duties totalled £2125 16s 11d during the time covered by the larger diary from September 1672 to December 1680. He also received payments directly from the City Authorities for his office of surveyor.

The greater number of the private dwellings had been set out by the time that he commenced his diary, so that the sum of money he had received for his viewing work would have been several times the amount stated above. He kept his gold and silver coins in a chest, which at his death contained several thousands of pounds.

The rebuilding programme also gave him the opportunity to design buildings, and he was the architect for the new Bethlehem Hospital and for the Royal College of Physicians. He was also engaged on work in rebuilding the City Companies' halls, and there are several references in the diary to the Grocers' Hall, Merchant Taylors' Hall and the Mercers' Hall. The entries in the diary just record visits and are brief. The records at the Mercers' Hall also show payments, but no detail of the work done. Hooke's many references to the Monument show that he was the designer as well as the surveyor of the building, and he very appropriately made its height 202 feet, the same measure as its distance from the baker's shop in Pudding Lane where the Fire began. He also had in mind its use for scientific purposes, to be used as a zenith telescope, and the flaming orb at the top had an opening large enough to admit a man. Platforms fitted inside the Monument prevent this view today. He had also fitted a zenith telescope through the roof of Gresham College, and by this was able to be the first person to observe a star in daylight.

On 16th May 1678 he also recorded in his diary that he tried the experiment of noting the height of the mercury in the barometer at ground level and then at the top of the pillar, and that it fell by about a one-third of an inch.

He also recorded on 10th April 1679, "At Fish Street Piller. Knight cut a wrong R for P."

Of the City churches St Magnus, St Brides, St Mary-le-bow, St Martin's and St Stephen's, Walbrook, have the most entries in his diary.

On 12th April 1675 the entry reads, "To Tompions and left at Sir Ch. Wrens the 14 ground plats of churches Harry had drawn. Missed him."

On 21st May of the same year, amongst other payments was "£150 from Sir Ch. Wren for church accounts", "and also from Mr Mountacue £100, from Hospital £100, from Physicians £20."

Two other large developments in the City are mentioned in the diary. The Fleet Canal was one that he had to supervise. This was a scheme to make the River Fleet navigable as far as Holborn Bridge, and to build wharves and quays along each side. This work was undertaken by Thomas Fitch and completed by 1675 at a cost of £51,000. The scheme was not a success and in 1733 the canal between Holborn Bridge and Fleet Bridge was arched over, and this was later extended to the bank of the Thames. John Evelyn had hoped that the ground from St Paul's to the Fleet would be left open, but this idea also did not come to fruition.

The other scheme which would have enhanced the river frontage was to build a Thames quay from the Tower to the Temple, but this also failed to materialize. Hooke's drawing for this plan survives.

The Great Fire considerably extended Hooke's activities, and like Wren, gave him the opportunity to develop an interest as an architect. It also enabled him to obtain commissions from a number of wealthy people. The diary records their names and the numerous meetings with Hooke. Research since the publication of the diary has confirmed these brief notes.

The Dictionary of British Architects lists twenty-four

buildings which he designed, or to which he made additions or alterations. In the City of London this includes the Haberdashers' Aske's Hospital, Hoxton, Bethlehem Hospital, Moorfields, Christ's Hospital, Newgate Street, The Writing School, Merchant Taylors' Hall, the Hall Screen, The Royal College of Physicians, and the Monument in association with Wren. Other works outside the City were a house at Spring Gardens for Sir Robert Southwell, five houses in the Strand for John Hervey, and two in St James Square for John Hervey and the 1st Earl of Ranelagh, and the stables at Somerset House in the Strand for Queen Catherine of Braganza. The largest private house built by Hooke was Montagu House.

Evelyn records visiting it on 11th May 1676. "I dined with Mr Charleton, and went to see Mr Montagu's new palace near Bloomsbury, built by Mr Hooke of our Society, after the French manner".

Celia Fiennes also compares the echo in one of the rooms there with that in Gloucester Cathedral. "It's the wall carrys the voyce – this seems not so wonderfull as I have heard for the large roome in Mountagu House (soe remarkable for fine painting) I have been in it and when the doores are shutt it's soe well suited in the walls you cannot tell where to find the doore if a stranger, and it's a large roome every way; I saw a Lady stand at one corner and turn her self to the wall and whisper'd, the voice came very cleer and plaine to the Company that stood at the crosse corner of the roome soe that it could not be carry'd by the side wall, it must be the arch overhead which was a great height."

The building was damaged by fire in 1685, and rebuilt by the French architect Puget.

Celia Fiennes visited it again between the years 1701–1712, and wrote, "Lord Mountagu's house indeed has been new built and is very fine; one roome in the middle of the building is of a surpriseing height curiously painted and very large, yet soe contrived that speake very low the wall or wanscote in one corner and it should be heard with advantage in the very opposite corner across – this I heard Myself."

This makes one wonder as to the extent of the fire in affecting the main structure in view of the remarks of this very observant traveller. The British Museum now occupies this site.

Hooke altered the design of Ragley Hall for Lord Conway. The entries in the diary from 15th June 1680 deal with his journey to Lord Conway's, his buying of a "camlet" coat at the Fox and Goose of Bull for which he paid 36s 6d, and also a whip for 2s and a pair of gloves 2s.

After riding to Ragley and making a few alterations to the design, he was given 30 guineas by Lord Conway. He took leave of him on the following day, gave 25s to the servants, and took horse at 10 a.m He lay very scurvily at Islip that night, and on the following day took horse at 4 a.m., reached Beaconsfield by 10, where he dined, and reached Gresham College by 6 p.m.

His companion Davys, "was seased with his ague. I was not in the least weary. Went with Society to Jonathons stayed till 10".

On the following day he visited Sir W. Jones, whose house in Bloomsbury he had visited on 17th April 1680, and which Hooke had designed or altered for him. Hooke designed and built Ramsbury Manor in Wiltshire for Sir William, and this has survived to the present day.

These buildings, together with Bishop Seth Ward's Almshouses at Buntingford, work at Magdalen College, Cambridge, panelling in the choir at Canterbury Cathedral, Escot House in Devonshire, are all listed in the Dictionary of British Architects as Hooke's work. The origins of these works have all been researched by historians of architecture.

There are other large building projects mentioned in the diary and for which there are several references. The building of Burlington House was originally commenced by Sir John Denham, of whom Evelyn said he was a better poet than architect. Sir John died in 1668, and from the number of entries in the diary the editors in a footnote expressed the opinion that Hooke completed the building. It was originally built in brick and later added to and cased in stone during the life of the second Earl. Entries in the diary relating to this are:

1st December 1676	"With Tom to Lady Burlington. She was much pleased with diagonall descents."
6th January 1677	"To Lady Burlington. Mett man and concluded garden."
16th January 1677	"Thence to Lady Burlington. Gave her an account of lead pipes. Talbot demanded £2 10s per lb., but I suppose he will take 20s per lb."
20th February 1677	"At Lady Burlington asked about £5000."
14th March 1677	"Made a draught for Lady Burlington."
23rd March 1677	"At Lady Burlingtons. 2 guineas about Porter's Lodge and peers."
5th June 1677	"Calld for draught at Burlington House."

There are several entries relating to Sir R. Edgecombe, for whom Hooke made a draft of a house on 30th July 1675, and a second draft on 2nd August, which he showed to Sir R. Edgecombe on the same day, in the company of Fitch, the builder engaged on the building of the Fleet Ditch.

On 1st September he went with a Mr Austin to view a house in Leicester Fields, and in Air Street, and where the entries end.

On 5th November 1677, Hooke also agreed a contract with Angiers for £4,800, after much bargaining.

Hooke in his diary had earlier referred to Angiers as "a Hudson's Bay Jesuite". There are no more entries giving any detail as to progress or completion of the contract.

There are several entries in the diary for work done for

Dr Nicholas Barbon, the son of "Praise God Barebone" after whom one of the Commonwealth Parliaments was named. Dr Barbon engaged in speculative building which was described in a paper read by N. G. Brett-James to the Society of Antiquaries on 29th March 1928. Dr Barbon is also mentioned in Marx's *Das Kapital* as the originator of some economic principles.

Hooke also drew a design of a house for Sir R. Reddings on 13th October 1675, and was also involved with him, Sir J. Hayes, Col Philips and Fitch in a scheme to build a mole at Tangier, which had been ceded to Britain by Portugal under the terms of Charles II's marriage to Catherine of Braganza. This scheme did not materialise.

One naval development which the *Dictionary of British Architects* suggests may possibly have been designed by Hooke were The Officer's Dwelling Houses, the Great Store, Rope House etc. at Plymouth Dockyard, as it was noted that a set of designs of these buildings, as engraved for the benefit of the naval Commissioners is among the Hooke papers in the British Museum, and the design closely resembles that of Bethlehem Hospital.

As with all subjects in which he took an interest, Hooke read up on them before he embarked on his own efforts. On 7th November 1674, he bought a book on architecture by Vingboon and Muet, from Thompson in Bedford Street for £1 10s. In the following January he purchased a book by Vitruvius from Martin for 8d, and in March of the same year another book by the same author from Oldenburg for £1 14s. In July he bought one on French architecture from Littlebury for 1s and in the following year he purchased a set of 90 pages of *Bachinall Grotesques*.

"Ceelings, gates, compartments and shields, beside the Palace of Richelieu, and the church of the Sorbonne at Large, at a cost of 15 shillings from Lap."

As France was the most powerful nation in Europe at the time, and Louis XIV was giving expression of that greatness in building great palaces, this was a suitable example to follow.

A NEW BUILDING FOR THE ROYAL SOCIETY

After the Fire the Society did not meet on its customary date, "by reason of the late dreadful fire in London". The council of the Society met on 12th September, and decided to hold their next full meeting at Dr Pope's lodgings in Gresham College. As the former rooms where they had met at Gresham College were now taken up by the Lord Mayor and his officials it was agreed to hold the next meeting at Arundel House. A year later, at the meeting on 30th September 1667, Dr Wilkins moved that a committee of both the Society and the council might consider raising contributions among the members in order to build a college.

The meeting on 5th November considered that the building of their own college would be the best way to establish the Society, and that the King might not be unwilling to give the land of the Wardrobe for erecting such a college. On 11th January Mr Hooke was desired to prepare a draft plan for such a building. The idea met with an enthusiastic response from some members. Sir Samuel Tuke was ready to pay his arrears, but also to contribute to the proposed new building, according to his abilities, and he thought the Earl of Devonshire and Mr Povey would also do their part. Dr Wilkins mentioned that Mr Nelthorp and Mr Skippon would contribute. Sir Anthony Morgan was asked to attend Mr Howard to consider with him the best way of securing to the Society his conveyance of the ground at Arundel House for building a college.

Two hundred letters were to be printed and sent to members asking for their subscriptions towards the new building. The President, Lord Brouncker, subscribed £100, and Mr James Hayes £40. Dr Wilkins subscribed £50. Sir

Robert Moray was to be asked to invite the nobility of Scotland who were members of the Royal Society to subscribe. The Lord Bishop of Salisbury gave £100. He was also asked to appeal to the other bishops for their support, and Henry Howard of Norfolk was to speak to the temporal lords for their contributions. Mr Colwall gave £100. The amanuensis was to make a book with leaves of vellum in which the names of contributors could be recorded. The president, Mr Henry Howard, and Dr Wilkins were to meet in Westminster Hall and solicit members who were also Members of Parliament for their subscriptions. By May 1668, £1000 had already been subscribed, and the president moved that the building of the college be begun forthwith.

Dr Wilkins was asked to obtain Dr Wren's draft for the building, and Mr Hooke was also asked for his draft, and his estimate of the cost. It was reported on 30th May 1668, that Mr Henry had set out the ground for building at Arundel House, a site 100 feet by 40 feet. Dr Wren was asked to meet Mr Henry Howard at Oxford about the draft of the building. Wren replied to Oldenburg, secretary of the Society, sending his details of his design and an estimate of the cost, £2000. On 22nd June Mr Hoskyns reported that he had received assurances from Mr Henry Howard about making the ground available for the Royal Society, notwithstanding there was an act of entail on it dating from the reign of Charles I. Hooke's draft of the building was agreed upon and Hooke was ordered to make a model of it with one door; to consider the buying of materials and contracting with workmen to do the work, "to be paid so much a rod and square", and also find a person who could be personally present to oversee the work.

On 29th June he was ordered to bring in at the next meeting of the council an estimate of the charge for materials and workmanship. A week later he was ordered to make another draft for the building of the college representing the front facing the Thames, and showing the arrangement of the windows. Hooke was again ordered to contract the workmen, look after the materials, and make an estimate of the charges. The treasurer was ordered to get in such portions of subscriptions as had been promised. The two drafts prepared by Hooke and Wren were examined on 13th July, both of which showed the building facing the waterfront, but the decision as to which should be followed was deferred until the next meeting, when Mr Howard's arrangements for the conveyance of the land to the Society should also be made known.

On 10th August it was resolved that the building of the college should be deferred until the spring, and in the meantime good materials should be provided. Mr Howard promised that he would endeavour to procure an Act of Parliament for letting of leases against that time.

The fact that there were no other mentions of this proposed building after August 1668 indicates that the scheme was allowed to lapse. One can only conjecture as to the reason. The money had been raised or promised, Henry Howard had agreed to obtain a lease of the land to the Royal Society, subject to receiving release from the restrictions of the act passed in the reign of Charles I. The failure to provide a permanent centre for the Society and its varied activities was regrettable as the accumulated collection of books, letters, papers and various curios given to the Society in its early years would have been of considerable interest, as would the many instruments made. The later move of the Society from Gresham College to Crane Court may have resulted in some of these being considered of insufficient value to be retained, but, like the exhibits in the old Ashmolean Museum, they would have been evidence of the interests and abilities of the original members, and a link to later developments.

Dr Wren's draft for the proposed building was described in a letter to Oldenburg, dated 7th June from Oxford.

Its length was 100 ft and breadth 30 ft. The great room for the meetings was to be 40 ft long, and two stories high, and the screen between the antechamber and the main room could be opened to enlarge the meeting space to 55 ft long. (p. 290–291 Birch Vol II). The estimated cost was £2000, and he was willing to prepare a model. "The cupola may be left till the finishing."

On 13th June 1678 it was reported that the Duke of Norfolk was having Arundel House pulled down, and Mr Hooke and Mr Hall were asked to attend his grace, and request that the library which he had bestowed on the Society could be removed to Gresham College, where room had been provided for it. They were also to deliver to the Duke the catalogue of the library which had been made by Hooke. It would appear from this action that the Duke had changed his mind about his willingness to provide the land.

BLOOD TRANSFUSION AND SKIN GRAFTING

Aubrey in his *Brief Lives* stated that Francis Potter had the idea of curing diseases by transfusing blood from one man to another, and that the hint of this came into his head when reflecting on the story by Ovid of Medea and Jason. Aubrey and Potter tried the experiment with two hens, but the birds were too small, and their tools not good. Aubrey then sent him a surgeon's lancet. He said that he had received a letter from Potter in 1652 about this subject, but later Dr Power claimed that he was the first person to attempt it. Wren also tried blood transfusion with chickens, and similar experiments were being made in France at that time.

At a meeting of the Society on 17th May 1665, the experiment of injecting Florentine oil of tobacco into the veins of a dog was made, but this did not cause any apparent change. Dr Wilkins then proposed that the experiment of injecting the blood from one dog into another should be tried. Lord Brereton mentioned that a horse tired after a long journey to London was given a drink of sheep or calf's blood which strengthened and revived it.

A week later Dr Wilkins, Mr Daniel Coxe, Mr Thomas Coxe and Mr Hooke were appointed to take care of injecting the blood of one dog into the veins of another, and Mr Thomas Coxe was particularly desired to try changing of dog's skins. Thomas Coxe also reported that he had tried injecting the blood from one pigeon into another, and this bird lived for half an hour after the operation.

Dr Wilkins reported the result of the experiment he had carried out, when five or six ounces of blood from one dog were let into a bladder, and then pumped into another dog. The first dog died, but the other lived.

There was then an interval from this date, 7th June 1665, until February of the following year due to the Plague.

When Fellows met in February 1666 they reported on their activities while out of London.

Sir Robert Moray mentioned that Robert Boyle had been making attempts at transfusion of blood from one animal to another. Dr Clarke stated that he had tried this, but had found it too difficult, and given up the attempt.

At the meeting on 20th June 1666, Dr Wallis reported the success of Dr Lower in injecting the blood of a mastiff into the jugular vein of a greyhound, using a quill. The mastiff died, having lost nearly all its blood, but the greyhound survived and ran away well. The report of this experiment was ordered to be registered, and Mr Daniel Coxe, Mr Thomas Coxe, Mr King and Mr Hooke were ordered to repeat the experiment in private, and again before the Society.

Mr King and Mr Thomas Coxe demonstrated this to the meeting of the Society on 14th November, transfusing the blood from a mastiff into a spaniel, with good success. The mastiff died, but the spaniel survived. Mr King gave a detailed description of this, and other similar experiments, firstly between two sheep, then between a small bulldog and a spaniel. In each case the recipient was very well and brisk. The next order was to perform the operation between a sheep and a mastiff, and from a young healthy dog into a sick one. In these experiments a brass pipe was used instead of quills.

Further suggested experiments were that old and sick animals should receive the blood of young, healthy ones and between different species, such as the blood of an ox or cow into a diseased horse. Mr Boyle moved that the animals should be weighed before the operation. Mr Thomas Coxe reported the success of transfusing the blood of a sound dog into a mangy one, which was cured by the operation. On 21st March 1668, Mr Oldenburg reported that similar operations were being attempted in Paris.

At the meeting of the Society on 4th April 1667, Dr King

reported how he had transfused the blood of a calf into a sheep, and also injected sugared milk into the veins of a dog. It was reported that the dog which had had sugared milk in its veins had died, and stunk strongly before it expired. The sheep was let out to graze after the experiment. Mr Boyle suggested allowing a dog to bleed to death, to measure how much blood it contained, and Sir George Ent proposed letting a dog bleed almost to death to see whether it could be revived by a blood transfusion.

Dr King reported his experiment of transfusing the blood of a dog into a sheep, which was sick afterwards, but better after some of its blood was let. It was thought that too much of the dog's blood had been given to it. Dr King also reported that the sheep which had received calf's blood, and then put out to grass, had died after three weeks. He was desired to repeat the experiment, and then send the sheep to graze in Mr Henshaw's garden at Kensington.

A week later Dr Wilkins gave a report to the meeting of two experiments on transfusion performed in Paris. The first was on a youth of fifteen or sixteen who had a fever, was lethargic, had lost his memory, and for whom there was little chance of life. He was first blooded, and then given the blood of a lamb. On the following day he got up at five, was lively and well and went about his business. The other experiment was on a labouring man, who first had ten ounces of blood taken from a vein, and then twice that amount of lamb's blood put in an artery. Afterwards he felt very well and lively, and said they could repeat the experiment on him as many times as they liked.

On 25th July, Dr King brought in reports of experiments which he had performed on a sheep and a dog, and a fox and a lamb. The fox died, and when he dissected it he found the thorax and abdomen half full of bloody water, which made him wonder whether the lamb's blood altered the quality and consistency of the fox's blood.

The next development was when Mr Oldenburg suggested that the transfusion of blood to men might be tried, as had been done in Paris. Mr Hooke was desired to speak to Dr Allen, the physician at Bedlam, to suggest that the experiment might be done on some mad person there. Dr Allen's first response was not favourable, so a meeting of several physicians in the Society was called to consider how the experiment might be tried.

On 21st November 1667, Dr Lower reported to the Society that Arthur Coga, an inmate of Bedlam, was ready to submit to the experiment for a guinea. This offer was accepted, and it was agreed to perform the operation on the 23rd of the month at Arundel House, and that Dr Lower and Dr King were to perform it. Mr Coga was a bachelor of divinity, having graduated from Cambridge University. He spoke Latin well when he was in company, "but his brain was a little too warm".

The transfusion was performed before an audience consisting of Mr Henry Howard, his two sons and the Bishop of Salisbury and several physicians. The patient said that he felt pretty well afterwards. Dr King reported that he was well and merry, drank a glass or two of Canary and smoked his pipe. He pulse was stronger and fuller than before, and during the day he passed two or three stools. He also said that he was very willing for the operation to be repeated. He was asked why he had not received the blood of some animal other than a sheep, to which he replied, '*Sanguis ovis symbolicam quandum faculatem habet cum sanguine Christi, quia Christus est agnus Dei.*'

A second transfusion of about fourteen ounces of sheep's blood was given to Mr Coga on 12th December, before a large crowd of people. A week later Mr Coga gave his account of the effects of the transfusion and said he felt very well. He had been a little feverish at first, but attributed that to drinking too much wine after the operation.

At the meeting in January 1668 Dr Croone reported that Dr Terne was willing to try the experiment on some morbid person at the hospital where he was physician. A week later Dr King brought to the meeting of the Royal Society the set of silver pipes ordered by Mr Townely for him to make the experiment of transfusion. In February Mr Oldenburg reported that he had received a letter from John Denis M.D. of Paris, giving details of a man cured of a phrensy by the transfusion of blood. Dr Clarke mentioned a poor distracted woman who might be a suitable person to try transfusion on, but as she had no means, the parish officers were to be spoken to to ensure her care after the operation.

There was an interval of more than a year before the next mention of transfusion when Oldenburg reported that several attempts made at Vienna included one on an old dog that could hardly move, but which was restored to health after receiving the blood of a younger dog.

The Bishop of Chester, Dr Wilkins, reported an unsuccessful attempt made by Dr Thruston in transfusing the blood of a sheep into a dog. The dog was weighed before the experiment and was 15 lb 3 oz. The sheep was allowed to bleed liberally into the dog and its weight increased to 17 lb. The dog showed signs of great pain and died. When it was dissected, its heart was found to be full of coagulated blood.

In the following January an attempt was made to feed a dog by blood alone which was injected into its veins every day. Physicians in the Society were asked to note this experiment, and to consider which patients in their care were most suitable for this treatment, and to suggest to governors of hospitals that they give permission for this.

The experiment of keeping one dog alive by injecting the blood of another dog into it regularly was mentioned by Dr Croone in March 1669, and he was requested to make the attempt with the assistance of Dr Allen and Mr Hooke. Dr Croone also suggested later an experiment to discover whether one animal could be kept alive without breathing by another animal doing so.

It was noted with regret, that experiments were being much practised abroad on transfusion and injection, and improvements made, but were neglected in England, where they were first invented. The lack of interest was shown by no further entries on the subject.

In his diary, Pepys recorded his conversation with Dr

Croone at the Pope's Head, when Dr Croone gave him an account of the blood transfusion experiments held at the Royal Society's meeting that day, 14th November 1666, when one Fellow posed the question as to what would be the result of transfusing the blood of a Quaker into an Archbishop. This question does not seem to have been registered in the transactions for that day, judging from the entries in Birch's *History of The Royal Society*.

The first report of skin grafting was on 21st October 1663, when the operator produced a dog, a piece of whose skin had been cut off and sewn on again. The operator related that as soon as the skin was cut off it shrank to half its size, so did not cover the whole area from which it had been cut. Mr Hooke was asked to give a written report of this.

On 4th May 1664, at the meeting of the Royal Society, it was ordered that Dr Croune, Dr Balle, and Mr Hooke take care at the next meeting to cut off some skin of a dog, and that the operator provide a dog for this purpose.

At the meeting on 18th May Dr Charlton cut off a piece of dog's skin to try whether it would grow on again. He reported at the next meeting that the dog had got the piece of skin off. He was asked to repeat the experiment, and think about a way of making the graft more secure. On 15th June a piece of skin was cut off from the neck of a dog, and stitched on again. The operator was instructed to take special care that the dog did not scratch it off. The operator at the next meeting reported that the dog had run away, and was ordered to provide another for the next meeting. Dr Wilkins and Dr Charlton were ordered to repeat this experiment, and take better care of the dog. This request was repeated on 6th July.

After this date interest seems to have lapsed. During the plague dogs were destroyed as being thought to be the carriers of the disease, so with the Society not meeting during the worst period of that, and then possibly a shortage of dogs after renewing their meetings, these facts might account for further experiments on the subject not being performed.

A UNIVERSAL LANGUAGE

A project of particular interest to Dr Wilkins was the devising of a universal language. John Comenius had dedicated his book *Via Lucis* to the Royal Society, and among the ideas proposed in it were schemes for a complete unification of knowledge, an international language, which would be a real character, and an organised scale of research and education to promote religious and social peace. These ideas were being developed when the Thirty Years War was still devastating parts of Europe. Theodore Haak, a German immigrant, championed these ideas in Britain, and Dr Wilkins was a staunch supporter of them.

The universal language for the Christian world before the Reformation had been Latin, but Luther, by translating the Bible into German, and before him, Wycliffe, Tyndale and Coverdale translating it into English, had made the word of God available to people in their own language. The Great Bible sponsored by Thomas Cromwell and Cranmer and fixed in all parish churches had aided the literacy of more people in their desire to read it, as later did the Authorised Bible of James I.

The great influence of this in England is the subject of Christopher Hill's book, *The English Bible, and the Seventeenth Century Revolution.*

The learning of the classical languages, Greek and Latin, was still the basic subject of the grammar and other charitable schools, a compulsion with which Milton in his *Essay on Education* disagreed. Scientific papers were still written in Latin, and William Gilbert's *De Magnete* was written in that language.

Although Hooke was a proficient Latin scholar, and enjoyed a conversation with Haak in that language, his first important book, *Micrographia* was written in English.

At the meeting of the Royal Society on 18th January

1665, "It was ordered that Dr Wilkins meet the first time (at least), with the committee for improving the English tongue; and that particularly he intimate to them the way of proceeding in that committee, according to the sense of the council, viz., chiefly to improve the philosophy of the language."

On 4th June 1666, after the meeting of the Society at Gresham College, Pepys returned to his house with Hooke, and lent him some books with details of naval matters, and the names of rigging and timbers used in shipbuilding to define these precisely for Dr Wilkins' book about the Universal Language, which was to be published soon. Dr Wilkins consulted John Ray and other scientists about the terms used in their respective subjects.

Dr Wilkins' *Essay Towards a Real Character and Philosophical Language* was read to the Royal Society on 14th May 1668, and a committee was appointed to read and consider the book. The committee was ordered to bring in this report on 29th October, but there is no record in Birch's *History of the Royal Society* of them doing so.

On 1st February 1669, Dr Wilkins proposed that Dr Sprat's *History of the Royal Society* should be translated into Latin, and that thirty pounds would satisfy him.

The reform of the English language to make it universal was a frequent topic of conversation, noted in Hooke's diary. On 21st December 1677, Hooke was with Sir John Hoskins at Jonathon's, and told him of his way of regulating it to make it a universal language, and Hill took a note of it in writing. A week later, on the 28th, at Jonathon's there was much discourse with Lodowick, Hill, etc., about reforming the English language.

Hooke went with Sir John Hoskins to Westminster Hall on 15th April 1678, where they met Mr Smith, Dean Lloyd and Sir Christopher Wren, and this was again the topic of conversation, as well as maps and a new shorthand. The last topic might apply to the example of the way of writing the universal language given which describes the basic principle of a watch.

With the passage of time, and three hundred and thirty years later, English has become a universal language, spread by colonisation and commerce, and by the rapid increase in the speed of communication of all kinds. As language is always developing as new inventions are made and new discoveries, the same object may have a different word, even in the English-speaking world. Sometimes politicians may seek to hide their actions by coining a new term, but genocide has meant the same thing since before Tamurlane butchered his captives, Hitler and Stalin did the same to those they despised, and the same actions continue today. In spite of the wealth of words in English there is a tendency to reduce the number of adjectives to two, 'positive' and 'negative', possibly a development from the binary system but, as Socrates often asked, 'What do you mean?'

In his doctoral thesis, *The Scientific Method and Mechanical Investigations of Robert Hooke*, Patri Pugliese of Harvard University wrote that Hooke thought that a clear, simple style of writing should be used, avoiding florid language, and should be concise. The entries in Hooke's diaries are models of brevity, unlike those of Pepys and Evelyn, and so require reference to his other writings and records of his experiments. The lectures given by Hooke and edited by Waller and Derham are detailed, thorough and interesting reading, although the precise meaning of some terms are distinctive to that age of scientific discovery, and may require reference to Dr Johnson's dictionary, as were some terms in Dr Wilkins' *Mathematical Magick*.

HISTORY OF THE WEATHER.

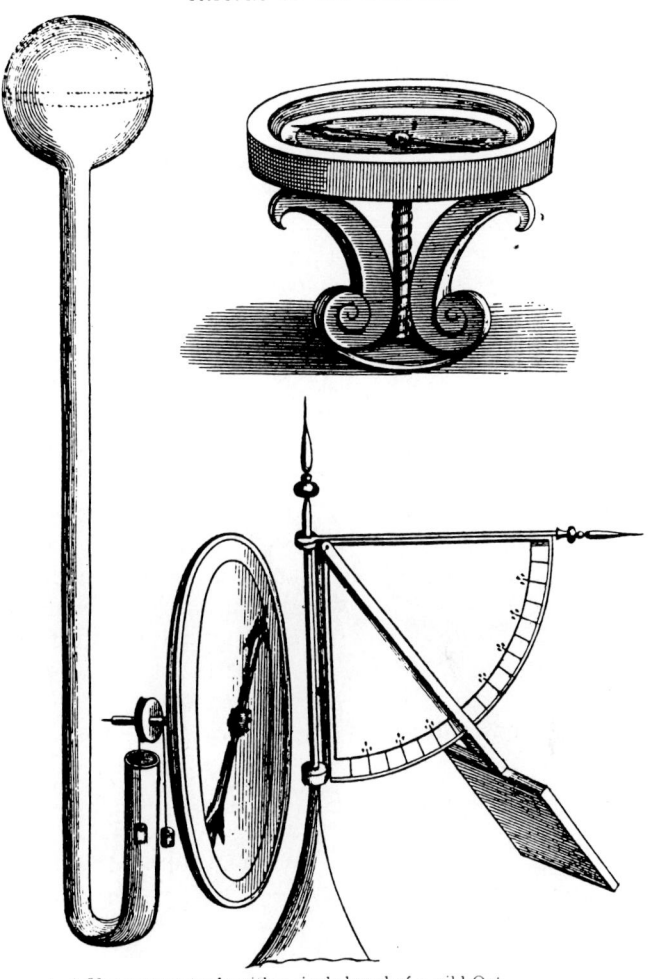

1. A Hygroscope made with a single beard of a wild Oat.
2. An Instrument with Quicksilver contrived with an Index to sensibly exhibit minute variations of Pressure in the Air.
3. Instrument for measuring the Strength of the Wind.
(Sprat, *History of the Royal Society*, 1667, p. 173.)

Wheel Barometer, Hygroscope and Anemometer – all used for recording details of weather conditions

HOOKE'S PNEUMATIC ENGINE OR AIR PUMP.
As constructed for Robert Boyle and used by him in his Laboratory in the High Street in Oxford.

Hooke's Pneumatic Engine or Air Pump

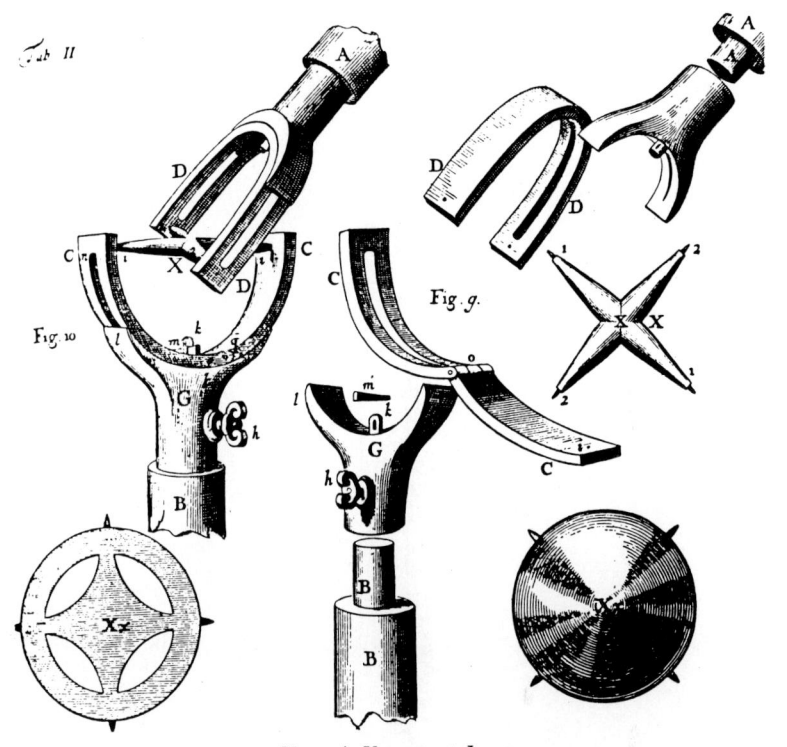

HOOKE'S UNIVERSAL JOINT.

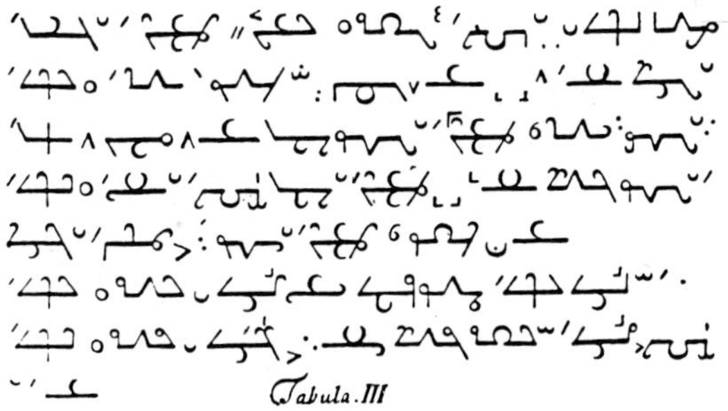

EXAMPLE OF THE UNIVERSAL AND REAL CHARACTER INVENTED BY
DR. WILKINS.

'A character and language perfectly free from all manner of ambiguity, ...
the most easie to be understood and learnt in the world.'

Hooke's Universal Joint

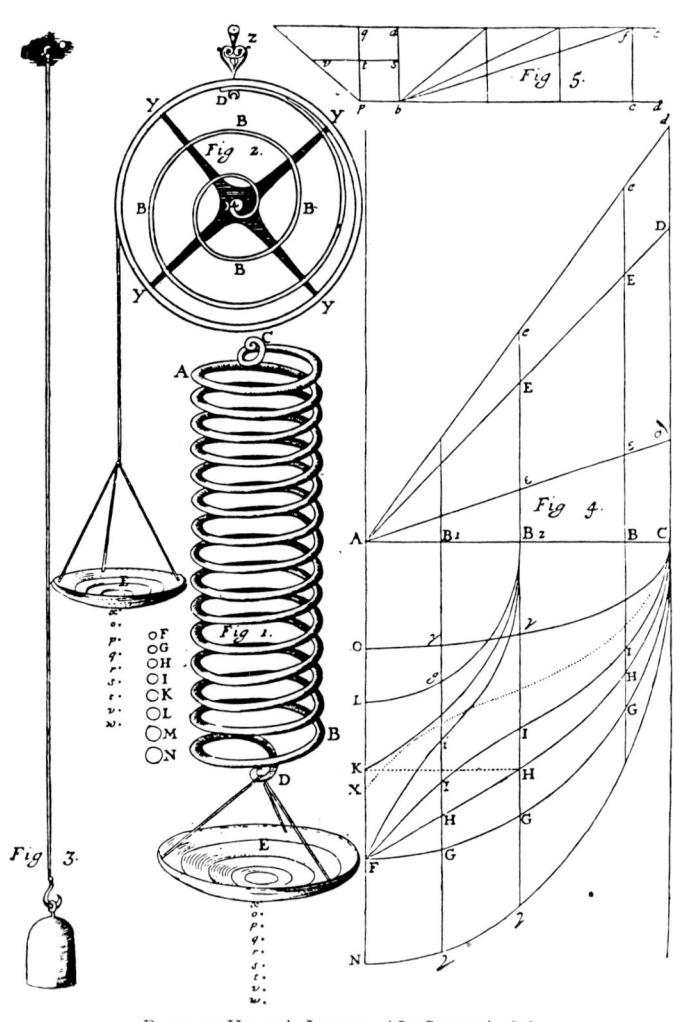

PLATE TO HOOKE'S LECTURE 'OF SPRING' 1678.

FIG. 1. Wire helical spring stretched to points s, p, q, r, s, t, v, w, by weights F, G, H, I, K, L, M, N.
FIG. 2. Watch spring similarly stretched by weights put in pan.
FIG. 3. The 'Springing of a string of Brass Wire 36 ft. long'.
FIG. 4. Diagram of velocities of springs.
FIG. 5. Diagram of law of ascent and descent of heavy bodies.

Plate to Hooke's Lecture "Of Spring" 1678

Bethlehem Hospital, Moorfields. Designed and built by Robert Hooke

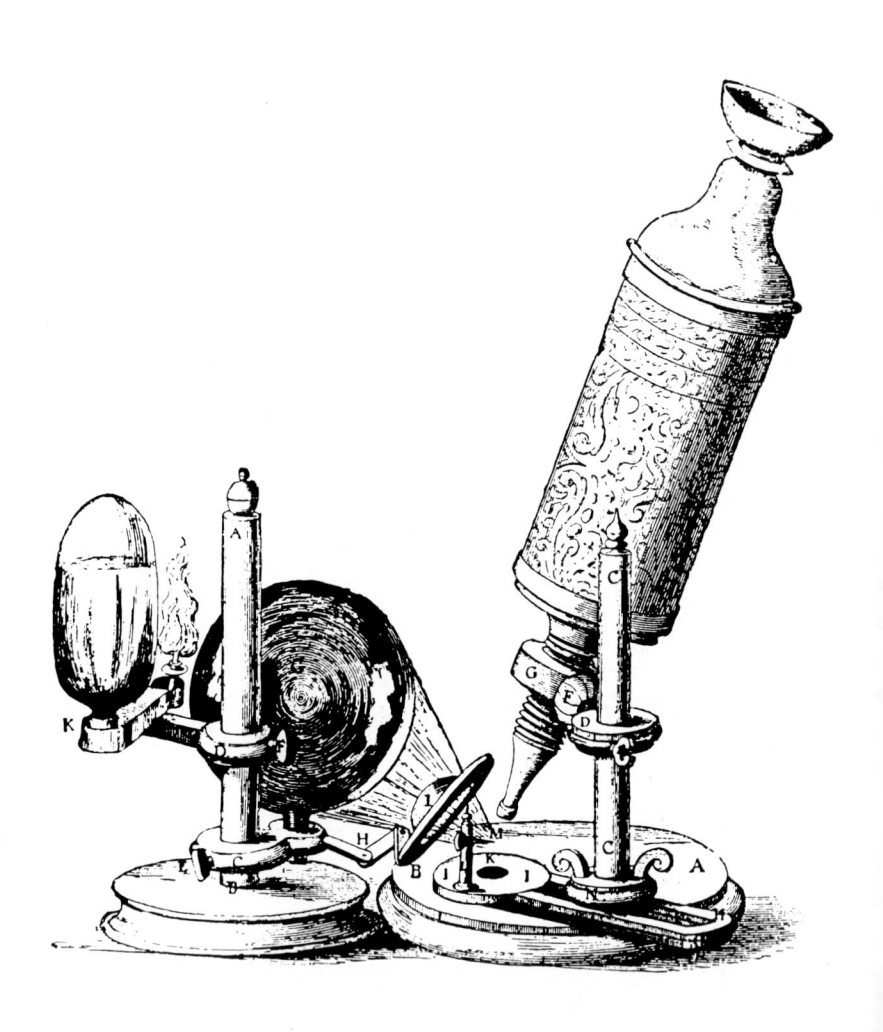

Hooke's Great Microscope, 1665

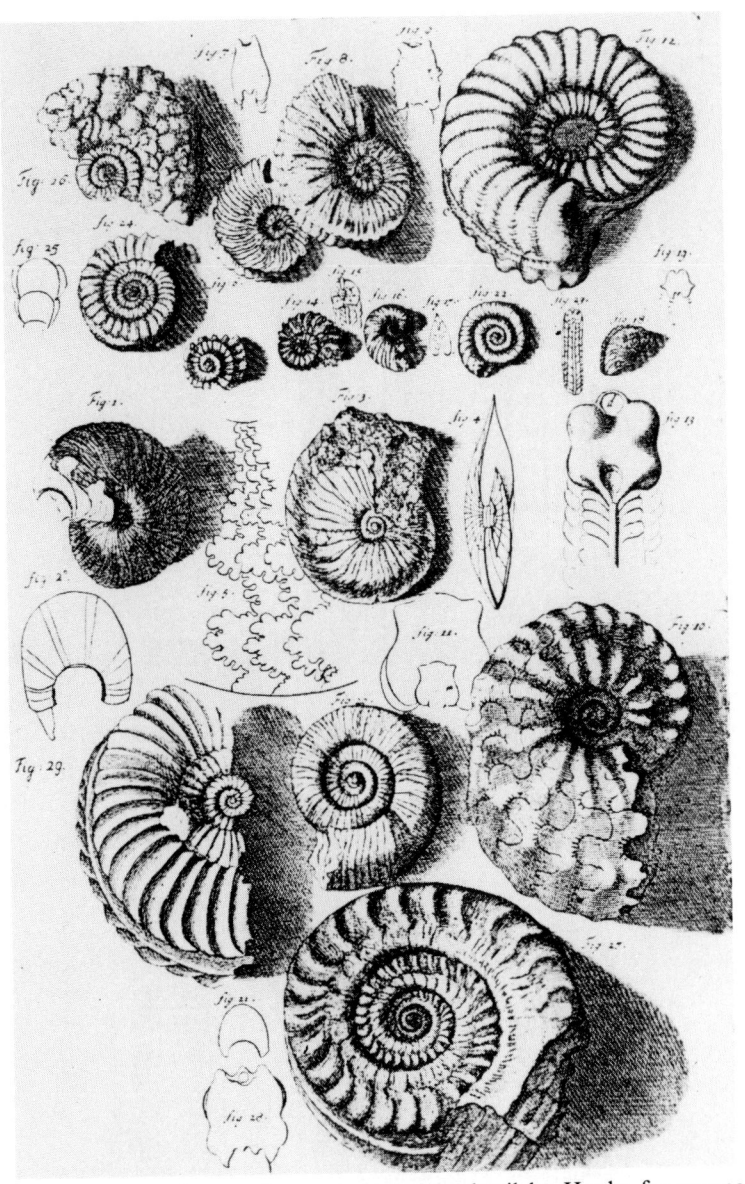

Snake-Stones or Cornua Ammonis drawn in detail by Hooke for comparison with recent Nautil-Shells

The Haberdashers Askes Hospital and School. Designed and built by Robert Hooke

The Monument, Fish Street Hill. Designed and built by Robert Hooke

n. d. Perhaps part of a Letter from HOOKE to Lord BROUNCKER.

Dr. HOOK's *Invention of a Reflecting Telescope.*[1]

I have lately made a telescope by reflection, with which I look directly at the object, and see it very distinct, and magnified. And this is by planting a small *lens* in the middle of the *object speculum*, and planting another small *concave speculum*, beyond the focus of the *object speculum*; the manner of which your Lordship will readily understand by the annexed scheme; where *a b* represents the object speculum, *e* the focus of that speculum,

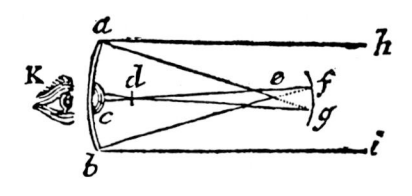

f g a small concave speculum, serving to reflect the rays to a second focus *d*, where the eye *k* see the object by the help of the small *lens c*. 'Tis easy so to contrive the cell for the eye, that the rays that pass on each side of *f g* shall not disturb vision.

We long much to hear of Mons. *Hugenius's* opticks and mechanicks. They are subjects capable of vast improvements, and cannot be rationally expected from any more likely, than from his acute wit and excellent pen. But, my Lord, I fear I have too far trespass'd upon your Lordship's patience, and must humbly therefore beg your Lordship's pardon, and subscribe my self,

     MY LORD,
      Your Lordship's most Faithful
       and most Humble Servant,
        R. HOOK.

Diagram of Reflecting Telescope. Designed and made by Robert Hooke

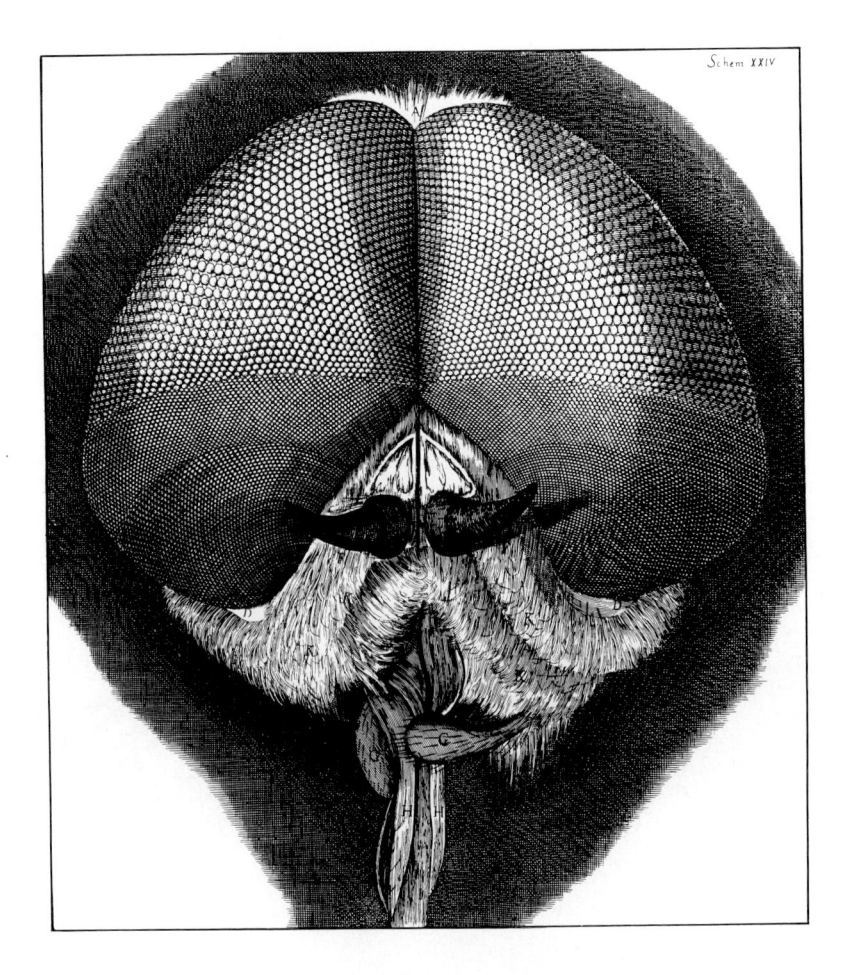

Eyes and Head of a Grey Drone Fly seen using Hooke's Great Microscope. Magnification 30

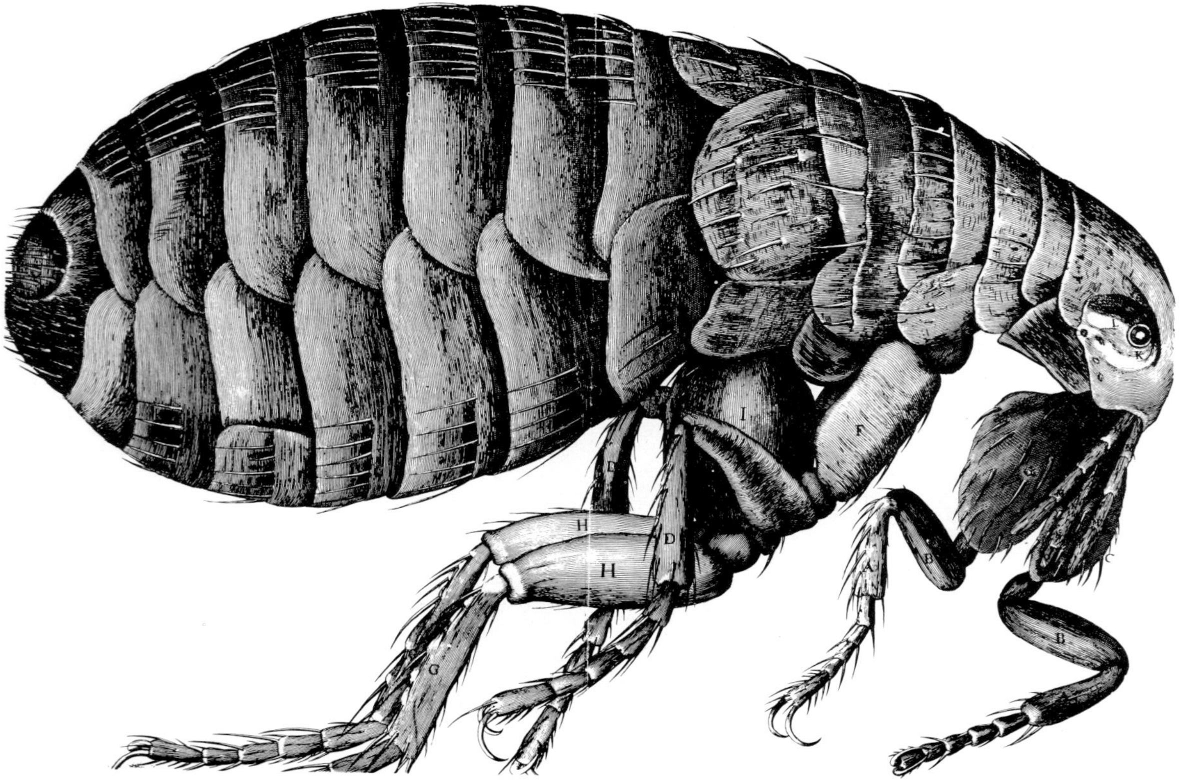

A Flea seen using Hooke's Great Microscope. Magnification 30

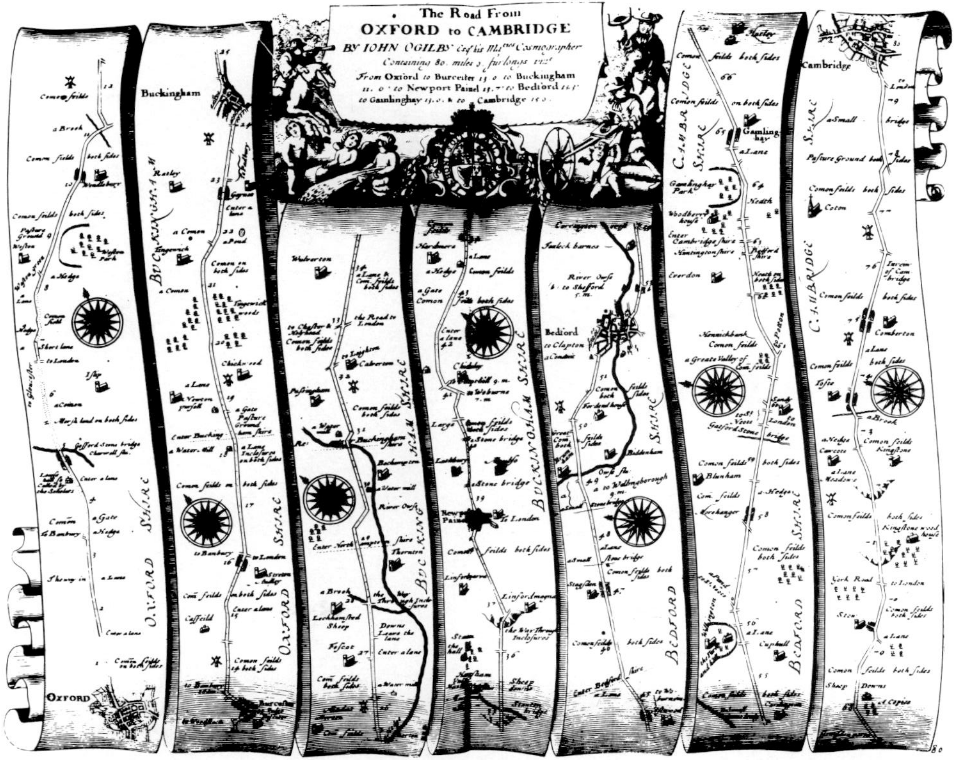

The Road from Cambridge to Oxford surveyed by John Olgiby using the Waywiser, which is illustrated near the title

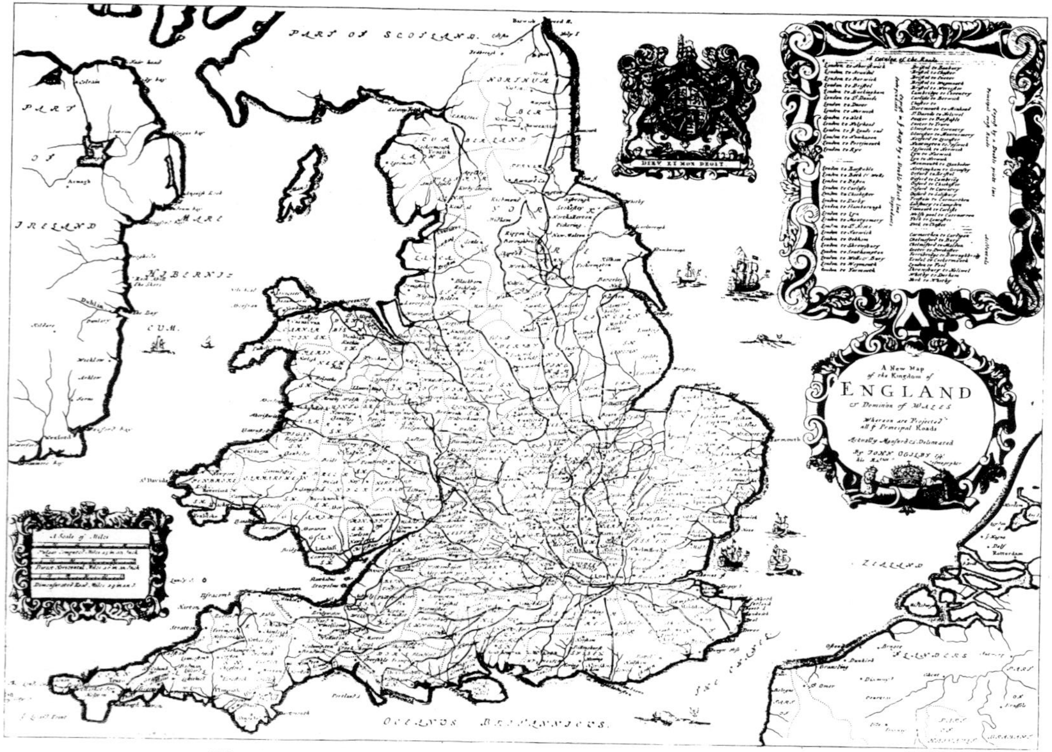

The Map of England and Wales made by John Olgiby, 1676

The North Prospect of MOUNTAGUE HOUSE.

Montagu House. Designed and built by Robert Hooke, 1676

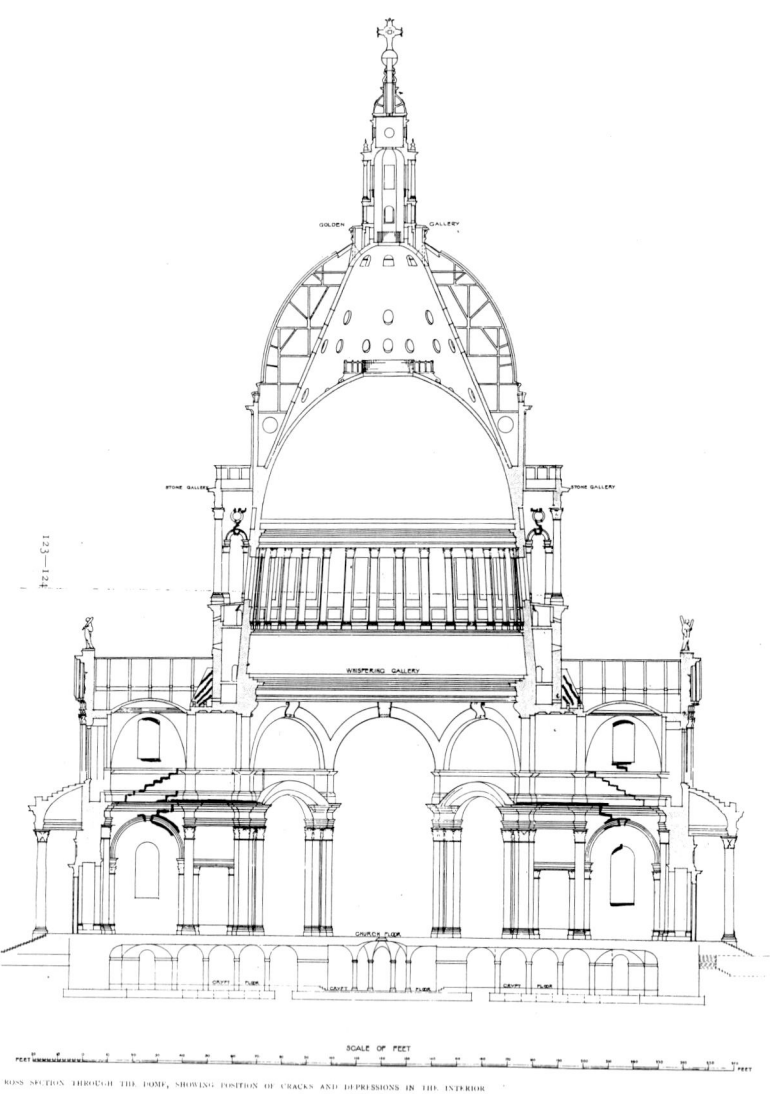

Construction of the Dome of St. Paul's built on the principles of the Catenary Curve and of Double Vaulting suggested by Hooke

HOOKE'S COLLEAGUES AND CONTEMPORARIES

JOHN AUBREY 1626–1697

John Aubrey was born at Easton Pierse, near Malmesbury, "on March 12th 1626, St Gregories Day, about the sun's rising, being very weak and likely to dye, that he was christened before morning prayers".

He was educated at home, which he regretted, as the house was in a large park, and he had no other children to talk to. His father employed a tutor for him, but Aubrey's greatest pleasure was in watching artisans, such as carpenters, masons and joiners who came to do work at the house. Later he went to school at Yatton Keynel, where the older boys learnt Latin from Virgil, Ovid and Cicero, and it was here that an older boy taught him his first lesson in morality by stealing his wooden top. In 1634 he moved to a school in the nearby parish, and during his stay there the school was visited by Thomas Hobbes, the philosopher, who spoke to Aubrey, and so began a friendship between the two which remained for the rest of their lives.

In 1638 Aubrey went to Blandford school, where two of his tormentors were the grandsons of Sir Walter Raleigh. Aubrey found it wise to make friendships with stronger boys, "as a line of protection".

In May 1642 he was entered as a commoner at Trinity College, Oxford, but his studies were disrupted by the outbreak of the Civil War, and his father ordered him to come home. He persuaded his father to let him return to Oxford after it had been garrisoned by the Royalist forces, but soon after his return there he was afflicted by smallpox, and after his recovery his father ordered him home again.

He always regretted not completing his studies, and his lack of any instruction in science.

He stated that his reading of *Religio Medici* in 1642 first opened his understanding, and from that time learning took a great leap forward. His great interest was in antiquities, and his concern was to preserve from further destruction the manuscripts, windows and monuments which had been damaged and scattered during the reign of Henry VIII and Edward VI at the time of the Reformation.

His first notable discovery was that of Avebury, and when Charles II heard of this he ordered Aubrey to accompany him on his journey to Bath, and show Avebury and Stonehenge to him. He walked up Silbury Hill and ordered Aubrey to bring him some of the very small snail shells found there. Charles also ordered him to make a detailed survey of Stonehenge, and during this Aubrey discovered the Aubrey Holes.

Earlier Aubrey had attended the meetings of the short-lived Rota Club, where he became acquainted with people who were later to be elected Fellows of the Royal Society, of which he was also elected a Fellow on 7th January 1663.

Aubrey stayed at Hooke's lodgings at Gresham College when he was in London and when he was on his journeys he used that address as his poste restante.

Aubrey's finances were not sound. His father had left debts of £1,800 at his death, and his estates were subject to lawsuits. Several entries in Hooke's diary are about Aubrey borrowing money from him, 20s on 14th October 1673, and another 20s on 25th November that year. On 20th November, Aubrey had settled the first debt by selling some of his books to Hooke. Aubrey was arrested for a debt of £200 on 5th March 1674, but four days later was released. Hooke noted "Aubrey cleared."

Aubrey continued to borrow money from Hooke during 1674, a few shillings at a time, and by 29th September 1674, this amounted to 45s. It looks as though this debt was settled by Aubrey selling his copy of Chrysostom to Hooke for £4 10s on 10th December that year.

There were occasions when they pitched upon founding a new club, as on 16th January 1674, and again on 11th December 1675 when they, and others, discussed Mr Newton's new hypothesis (of light). This meeting was at Joe's (Jonathon's).

On 18th December 1675, they met at the Garaway's club, and discoursed about, "the Universal Character, about preadmits and Creation. About insects".

Hooke mentioned that all vegetables were female. He also told Wild and Aubrey about flying. After a meeting of this same club on 1st January 1676, in the Green Room at Garaways, they then went by coach to Childs, where they stayed till 11 p.m., and when they returned to Gresham College, they found that they had been locked out. Hooke also mentioned a Rosicrucian club to Aubrey on 14th July 1676, but there were no other mentions of this.

Aubrey, Hooke, and Sir John Hoskyns went to the Palgrave's Head on 20th October 1676, where they were entertained to a meal by Sir Christopher Wren, whose birthday it was, and who paid for all. Hooke sent a hobby horse to young Wren the next day, which cost him 14s.

Aubrey and Hooke shared in some scientific activities. On 6th June 1674, "Mr Aubrey and I observed the Resistance of air to be duplicate to the velocity, or rather in a musicall proportion."

On 9th July 1674 Hooke showed his new quadrant to Sir Ch. Wren, Dr Goddard, Mr Hoskins, Dr Grew, Mr Aubrey and Mr Haak, but not to Oldenburg. Hooke attended a committee meeting of the Royal Society till 9.30 at night on 15th December 1675, *Libera me Domine*, was his comment. He then discoursed with Aubrey till 11.30 at night about a new mechanical principle of flying. On 1st June 1676, Hooke observed an eclipse with Aubrey, and in 1677, on 24th April, they looked out for the comet, along with Wild, Merrit and Moxon but missed it.

Dr Plot showed Dr Whistler, Mr Godfrey, Mr Hewks and Aubrey and Hooke some rushes and petrified mud on 6th October 1677, and on the same day Aubrey, "was pictured by Cooper". In the same month Aubrey canvassed for Hooke's election as Secretary of the Royal Society, and for Sir Joseph Williamson as President. Both were elected, but Hooke noted that Croon was in a great huff, while Colwell was merry at the results.

Hooke observed an eclipse of the moon on 19th October 1678 with Aubrey.

Two years later, on 5th October 1680, Hooke wrote that Aubrey was, "impudent", but did not give any circumstance or cause.

Aubrey wrote a book, *Ideas of Education of Young Gentlemen*, based on his own experiences at school, and his later reading of works by Samuel Hartlib, John Milton, Thomas Hobbes and Sir William Petty, all of whom advocated new methods, aims and curricula in educating the young. The only mention of education in Hooke's diary before the publishing of Aubrey's book in 1680, was on 17th November 1676, when Hooke, Hill, Hoskins, Wren, Gale and Aubrey discoursed about teaching children grammar by tables.

Aubrey's studies were not helped by his father, whose interests were solely in hunting and hawking. Aubrey wrote that his own studies were on horseback, or in the house of office.

"Till about the yeare 1649, when the New Experimental Philosophy was first cultivated at by a Club at Oxford, 'twas held a strange presumption for a man to attempt an innovation in learning; and not to be good manners, to be more knowing than his neighbours and forefathers; even to attempt an improvement in husbandry (though it succeeded with profit) was looked upon with an ill eye. Their neighbours did scorn to follow it, though not to do it was to their own detriment. 'Twas held a sin to make a scrutiny into the ways of Nature; whereas it is certainly a profound part of religion to glorify God in his Workes; and to take no notice at all of what is daily offered before our eyes is gross stupidity."

On 16th June 1674, Hooke went with Aubrey to see Hobbes, and saw him in Scotland Yard. Hooke had bought Hobbes *Thucidides* off Brook for 9s on 27th November 1672, and later lent it to Mr Haak, and to Mr Godfrey.

On 29th December 1678, Hooke heard Whitaker dispute against Hobbes. Hobbes like Hooke, took frequent exercise, walking rapidly until he sweated, and playing tennis three times a year, all done to keep him in good health. As he lived to the age of 93 his regular exercise was effective. As Aubrey noted, in his old age Hobbes was bald, and his greatest trouble was to keep flies from pitching on the baldness.

Pepys had bought a first edition of Hobbes *Leviathon* for 39s and on 3rd September 1668, noted that the price had risen to this amount because the Bishops would not let it be printed again, as it was considered atheistical. Like Bacon he was sceptical about the teaching of the texts of Aristotle in the universities of Christendom, exampled by his long chapter XXXIII, "on a Christian Commonwealth", in which he questioned much about the doctrine of the church, but he said in his conclusion that there was nothing in his discourse which was contrary either to the Word of God, or good manners.

Christopher Hill in his book, *The English Bible and the Seventeenth Century Revolution* quotes Dr Wilkins as defending the Copernican system, and that nevertheless, "astronomy proves God and Providence, and confirms the truth of the Bible".

There were those, such as Mr Henry Stubbe, the physician of Warwick, who charged the Royal Society with bringing into contempt the philosophy of Aristotle, undermining the universities, destroying the established religion, and introducing popery in its stead. This could not be said of either Hooke, Boyle or Evelyn, whose writings show that they were keen to arrive at a greater understanding of the works of the Creator, and appreciation of His omnipotence.

SIR WILLIAM PETTY 1623–1687

William Petty was the son of a clothier at Romsey and, like Aubrey, he was extremely interested in watching craftsmen at work; smiths, joiners, watchmakers and carpenters, until, eventually, he could work at any of their trades.

He became competent in Latin and Greek, and at the age of thirteen went to sea. He had the misfortune to break his leg, and was put ashore on the French coast, and left to fend for himself. This was near to the town of Caen, where he met some Jesuit priests who were impressed with his knowledge of Latin and admitted him as a pupil at their college. Here he learnt to speak French fluently, and also supported himself by giving lessons in navigation and in English.

He returned to England, but in 1643 went to the Netherlands, where he studied medicine at Utrecht and Leyden. In 1645 he went to Paris and continued his studies in medicine, reading Vesalius with Hobbes, and making the acquaintance of Father Mersenne. He returned to Romsey in 1646, and may have continued with his father's business of clothier. He also occupied himself with perfecting his system of double writing. In 1648 he moved to Oxford, where the organisation of the university had been reformed, and became Professor of Anatomy. He became a doctor of medicine and Fellow of Brasenose, Vice-Principal of that college, and Professor of Music at Gresham College.

He became famous when he revived a woman, Ann Green, who had been hanged for the murder of her child, taken down, and brought to the anatomy school for dissection. She later married and had a family.

In the early 1650s he was appointed Physician to the Army in Ireland, and during his stay there made the *Down Survey* of that country, which established both his fame and his fortune. He also became a firm friend of Henry

Cromwell, and it was as his friend that he was elected Member for West Looe in Richard Cromwell's parliament. In spite of his friendship with Henry Cromwell, he was knighted in 1661.

He was one of the founding Fellows of the Royal Society, and one of its most active, while in England. His interest in shipping led him to design a double bottomed boat, a form of catamaran, which proved to be seaworthy and faster than the Holyhead packet, the fastest of the King's ships. During the Great Plague, he, Dr Wilkins and Hooke stayed at Durdans, near Epsom.

It was there that Evelyn saw them on 4th August 1665, and wrote in his diary, "On my return I called at Durdans, where I found Dr Wilkins, Sir William Petty, and Mr Hooke, contriving chariots, new rigging for ships, a wheel for one to run races in, and other mechanical inventions; perhaps three such persons together were not to be found elsewhere in Europe for parts and ingenuity."

Sir William Petty applied the principles of matter, motion and number to his consideration of problems, and was one of the founders of the study of statistics, and in his book *Political Economy*, advocated the idea of the division of labour, which, as described in Tompion's organisation of his workshop, enabled Tompion to make thousands of watches and clocks to a high degree of accuracy.

Manuscript copies of Petty's *Political Arithmetick* were circulated among his friends, and Hooke wrote on 9th October 1673 that he gave Sir W. Petty's book to Hartlib, but did not give the title. On 13th October 1678 Hooke gave Dr Wood Sir W. Petty's paper, no title given, and he lent Hooke Petty's *Political Arithmetick*, and on 21st November that year, Hooke lent the book to Sir John Hoskins. On 12th December 1678, Dr Wood lent Hooke Sir Wm. Petty's *Book of Shipping*. According to the note, in Petty's writings, he finished the manuscript in 1676, and passed it to his friends for their comments before publishing it. He gave a copy to the King, but an authentic copy was not published until after his death.

As mentioned in Burton's diary, he was in favour of founding a university in London, so that the children of any people in humble circumstances could benefit from a good education, as he had done, and rise in the world.

DENIS PAPIN

The first mention of Denis Papin was on 3rd April 1676, when Hooke went to Boyles, "saw Slayer his room and pump, and Papin's engine". Hooke saw them all again on 19th June that year. The next mention was on 3rd February 1677, when Hooke spoke with Slayer about Papin's mending wind gun. On 19th May 1670 Hooke had produced an engine that might serve for a wind gun, but the valve was not yet ready, and he was ordered to have it ready for the next meeting. There was no mention of it at the next four meetings, and at the meeting on 16th February 1671, it was observed that very many things were begun, but few of them prosecuted.

Papin showed his wind gun to the Royal Society on 4th October 1677, and Hooke gave a detailed description of it in his diary for that date. Hooke helped Papin translate his description of his fountain on 10th April 1678, and on 13th June of that year, "M. Papin shewd the French designe of Employing cripples."

Hooke discovered that Papin knew of the rarefaction of gunpowder at a discussion on 9th July, and the use of it.

On 4th September Hooke called on Papin at dinner, and saw his wind gun and fountains. Hooke gave a compressing engine to Papin on 5th September, and called on him again on the 6th, and young Papin visited Hooke on the 8th. They met occasionally during the next few months, and on 5th January 1679, Hooke paid Papin 25s for his condensing engine and gave him a copy of his *Book of Springs*.

Papin's most notable invention, and one which caused great merriment at a meeting of the Royal Society when he demonstrated it was his "Digester"or pressure cooker. On Sunday 16th March 1679, Hooke watched this device in action softening bones, horns, etc., and on Sunday 27th April, Hooke watched Papin's digester softening bones and beef.

Three years later, on 12th April 1682, John Evelyn went with several Fellows of the Royal Society to a supper which, "was all dressed, both fish and flesh, in Monsieur Papin's digestors, by which the hardest bones of beef itself, and mutton, were made as soft as cheese, without water or other liquor, and with less than eight ounces of coals, producing an incredible quantity of gravy; and for close of all, a jelly made of the bones of beef, the best for clearness and good relish, and the most delicious that I have ever seen or tasted. We eat pike and other fish bones without impediment; but nothing exceeded the pigeons, which tasted just as if baked in a pie, all these being stewed in their own juice, without any addition of water. This philosophical supper caused much mirth and merriment amongst us, and exceedingly pleased all the company. I sent a glass of jelly to my wife, to the reproach of all that the ladies ever made of hartshorn."

ISAAC NEWTON

In the first diary there are thirty-four entries where Newton is named. The first was on 18th February 1675, and read, "Mr Newton, Cambridge here. Mr Newton told me his way of polishing metal on pitch."

On 21st September 1675, Hooke bought Newton's *Helps to Calculations* at Widow Page's shop in Fish Street Hill. Newton's paper on light was read to the Royal Society on 9th December 1675, and was the subject of the discourse. It was reported fully in Birch's *History of the Royal Society*, Vol. III pp. 247–260, and 261–270.

Two days later Hooke and others met at Joe's coffee house, and discoursed about Newton's paper. Newton's paper was read again at the meeting of the Royal Society on 16th December 1675, and was to be the subject discussed at the meeting of the new club on 1st and 20th January 1676. At the first meeting a letter of Newton's was read seeming to quarrel with Oldenburg's false suggestions, and Hooke told the meeting of his experiments about the colours of bodies about the dark spot in the middle about the visibleness of glass.

Hooke wrote in his diary, "I discoursed about the puls of light of a swash puls and a compounded puls." The remainder of Newton's paper was read on 10th February. Hooke had been asked to give his opinion on Newton's paper, and stated his criticisms. Pardies and Huyghens also had made criticisms of the paper.

On 24th November 1679, Hooke wrote to Newton suggesting that they engage in a private philosophical correspondence on scientific topics of mutual interest. Hooke invited Newton to comment on any of Hooke's hypotheses or opinions on the notion of, "compounding the celestial motions of the planetts out of a direct motion by a tangent and an attractive motion towards the central body".

On 23rd May 1666, Hooke had read a paper to the Royal Society in which he stated that all bodies having one impulse would move in a straight line, unless acted on by some other impulse, or impeded in some way.

All celestial bodies move in circular or elliptical lines, and so must be acted on by some other force than the first impulse. This might be the unequal density of the medium through which they moved. But the second possible cause could be from an attractive body placed in the centre, whereby it continually endeavoured to draw or attract that object to itself.

Hooke demonstrated this idea by swinging a ball of lignum vitae held by a string to the ceiling of the room. When the impetus at a tangent was stronger than the endeavour holding the ball to the centre, the orbit was elliptical, with the longest diameter parallel to the tangent. When the force to the centre was the stronger, the longest diameter was parallel to the line from the ball to the centre. When both forces were equal, the orbit was a circle.

"By this hypothesis, the phaenomenon of the comets and the planets may be solved." Birch, *History of the Royal Society*, Vol. 2, pp. 90–92.

These ideas were part of Hooke's Cutlerian Lecture, composed and read at Gresham College in 1670, and printed in *Early Science at Oxford* Vol. 8, pp. 27–8, and is possibly that referred to by Aubrey in his *Brief Life of Hooke* and quoted by R.S. Westfall in his biography of Newton, *Never at Rest* p. 382.

The inverse square law as it related to gravity, could have been deduced much earlier, as H.T. Pledge stated in *Science Since 1500*, by Boulliau in 1645, and by Wren and Hooke, and later by Halley. It was also applied by Horrocks in his calculations on the lunar orbits.

R.S. Westfall in his biography of Newton, pp. 382–3 quoted Hooke's *Attempt to Prove the Motion of the Earth* (1674, republished in *Hooke's Lectures* in 1679) but Westfall did not mention Hooke's earlier paper read to the Royal Society in May 1666, concerning the inflection of a direct motion into a curve by a supervening attractive principle, which lecture was ordered to be registered. Westfall states that the most remarkable aspect of Hooke's statement was that it correctly defined for the first time the dynamic elements of orbital motion, and that it was not an inconsiderable lesson for Newton to learn. I. Bernhard Cohen, in a paper on Newton's discovery of gravity, also stated that before this he had not reached Hooke's level of understanding of circular motion. He still often spoke of orbital motion in terms of centrifugal force.

Briefly Hooke's paper of 23rd May 1666, as reported in Birch's *History*, makes these statements:

1. A body once put in motion would move in a straight line, unless acted on by another force.
2. The second most probable cause of inflecting a direct motion into a curve may be from an attractive property of a body placed in the centre.

In another lecture on the collision of bodies, Hooke suspended three balls of lignum vitae in line, and allowed one to fall against the middle one, which apparently remained stationary, and the third ball swung outwards, and this motion continued. With four more similar balls suspended in line the demonstration is now known as "Newton's Cradle".

Hooke's annoyance at the fact that his contribution to Newton's work was not acknowledged in the first edition of the *Principia* was the reason for his comment on 3rd February 1689, when he dined at Dr Busby's.

Dr Hickman was there and said that Newton was, "the veryest knave in all the house".

On 15th February 1689, Hooke met Newton at Halley's where Newton, "vainly pretended claim yet acknowledged my information: Interest has no conscience: *A posse ad esse non valet consequentia*".

The only other mention of Newton in the second diary was on 3rd July 1689, when Hooke wrote, "Royal Society met: Hoskins Henshaw, Hill, Hall: dispute about Newton, of Leibnitz *fallere fallentens*. Newton and Mr Hamden came in, I went out. Returned not till 7."

This could be construed as Hooke not wishing to get involved in an argument between the two inventors of calculus, or to avoid meeting Newton.

Dr Hooke was mentioned in the second edition of the *Principia* published in 1713, ten years after his death.

Stephen Hawking in his book, *A Brief History of Time*, states of Newton that he was not a pleasant man. His relations with other academics were notorious, with most of his later life spent embroiled in heated disputes. The Rev Bailey described Newton's quarrel with Flamstead in his life of the first Astronomer Royal or, as he was first called, the Royal Observator. Newton was in a hurry to have some of Flamstead's observations, and Flamstead was busy getting in the harvest at his parish at Burstow in Surrey and, as the tithes were an important part of his income, he was concerned that all was safely gathered in. According to Stephen Hawking, Newton was incensed by Flamstead's action in taking Newton to court, and deleted all reference to Flamstead in later editions of the *Principia*.

That Hooke was of a more forgiving nature is evident from his association with Tompion, calling him a slug one day, and threatened to have nothing more to do with him, but a few days later called on him, and continued a fruitful partnership with him in the making of clocks, watches, barometers and other instruments into the reign of William and Mary. He also freely discussed ways of improving all these things with Tompion, all of which are dealt with in detail in R.W. Symonds biography of Thomas Tompion.

A. Rupert Hall in his book, *The Revolution In Science 1500–1750*, states on page 303 that by 1685 there is ample evidence that Hooke had a very complete picture of a mechanical system of the universe founded on universal gravitation. He also states that Hooke claimed that he had expounded these ideas as early as 1670. Hall, like Westfall, does not mention Hooke's paper read to the Royal Society on May 23rd 1666, when he first expounded his theory, and demonstrated it by swinging a ball of lignum vitae on a string. This was registered in the *Transactions of the Royal Society*, and quoted by Birch in his *History of the Royal Society*. In his reference to Hooke, reading a paper to the Royal Society in 1666, the date is not complete, neither are the details of the paper.

On 9th and 16th May 1666, Dr Wallis had stated his hypothesis that the earth and moon had a common centre of gravity. On page 304 of the same book, Rupert Hall states that Newton bitterly acknowledged later that much more was needed than mathematical expertise in order that a true mechanics could have been derived from Hooke's hints.

Newton could concentrate his efforts on his researches at Cambridge, with no other duties or distractions than the lectures that he gave to students. Hooke was at the beck and call of the Fellows of the Royal Society, investigating their queries, and expected to produce some of his own researches. Before the Great Fire he had his twice weekly lectures to give as the Gresham Professor of Geometry, and his Cutlerian lectures on the mechanical arts, both of which sets of lectures he continued to give. After the Great Fire he had the added duties as one of the three surveyors appointed by the City of London to supervise its rebuilding.

Rupert Hall stated, "Of all the early Fellows of the Royal Society his was the mind most sparkling imaginative. Schemes for new experiments and observations occurred to him so freely that each day was divided between a multiplicity of investigations, each in rapid succession subjected to his ingenuity and insight. He had a view – often it must be said, a self regarding view – on every topic raised at the Royal Society's meetings."

As, on reading Birch's *History of the Royal Society* one can see how often he was asked for his opinion, this is not remarkable. Rupert Hall also states that Hooke's first large work, his *Micrographia*, was his last, a statement disproved by the variety of work described in the diary of 1672–1680, and by the publication of *The Posthumous Works of Robert Hooke*, which contain his lectures on light, comets, earthquakes, navigation and astronomy, and illustrations of the scientific instruments which he designed. On the same page Rupert Hall also notes Hooke's Law of Springs, *Ut tensio sic vis*, which is an example of an action having an equal and opposite reaction.

Hooke's principle of the catenary curve, *Pendet continuum flexile, sic stabit grund rigidum*, was stated on 26th August 1675, and was followed by Wren in building the dome of St Paul's.

In his article about Hooke in volume X of *Pepys' Diary*, Hall refers to Hooke as the Operator of the Royal Society. Hooke was appointed Curator on 27th July 1664, a post which he held until his death on 3rd March 1703, and in which office, in the opinion of Marjorie Hope Nicholson, "he *was* the Royal Society".

It was as Curator that he was invited to dine with John Evelyn on 4th March 1664, and introduced as such to the Earl of Lauderdale, the Earl of Teviot, my Lord Viscount Brouncker, Dr Wilkins and Sir Robert Moray, the other guests.

The achievements of Sir Isaac Newton are summed up by Rupert Hall in the words, "Only an intellect of supreme clarity and percipience could erect a sharply defined, classical edifice from the jumbled materials available to mathematical physics in the 1660s."

As another famous Englishman of a slightly earlier date said, "Paint me warts and all".

A quotation from J. M. Keynes' speech on "Newton, the Man" given at the Royal Society on the occasion of the Newton Tercentenary Celebrations, 1946, stated, "in vulgar modern terms, Newton was profoundly neurotic of a not unfamiliar type, but – I should say from the records – a most extreme example. His deepest instincts were occult, esoteric, semantic – with profound shrinking from the world, a paralyzing fear of exposing his thoughts, his beliefs, his discoveries in all nakedness to the inspection and criticism of the world".

Another assessment from a review of *A Portrait of Isaac Newton* by F.E. Manuel in the Sunday Times of 16th February 1969, stated:

"He was a man of fearful rages and implacable in offence. He broke promises and withheld knowledge in order to advance his own glory. His quarrels with Hooke, himself a man of genius, and with Leibnitz, who was his true peer, are among the ugliest in intellectual history."

R.S. Westfall gives the picture of a man devoted to the pursuit of knowledge. His status at Trinity was one of isolation. He seldom left his chamber, preferred to dine there alone, and when he dined in hall ate there alone. He was hardly a genial companion, sitting silently, never starting a conversation. He did not join the Fellows on the bowling green. Of the three people who used to visit him, he ceased to have anything more to do with Vigani, because he told a loose story about a nun. None of this detracts from his genius or achievements, but as an isolated example, could account for his impatience with others.

SIR CHRISTOPHER WREN

There are more than 800 entries in Hooke's first diary relating to Wren, more than for any other person. The subject matter of these entries shows the wide scope of Hooke's duties and interest. His duties had increased with his appointment as one of the three surveyors for the City authorities. The author of the brief sketch of Hooke's life in the published diary of 1672–1680 stated that this period was one of the most active and productive of his life. In 1674 he published his *Animadversions against Hevelius*, in 1676 his paper on helioscopes, and in 1677 his *Lampas*.

Following the death of Oldenburg he was appointed Secretary of the Royal Society, and from 1679–1682 he published his *Philosophical Collections* in place of Oldenberg's *Philosophical Transactions*. In 1678 Hooke published his *Potentia Restitutiva*, which is said to contain some remarkably close anticipations of modern theories of elasticity and the kinetic theory of gases. Many of these ideas he discussed with Wren, as well as the many relating to the rebuilding of churches and other public buildings.

The entry for 2nd November 1672 records that Hooke dined with Wren and, "saw model of St. Paul's approved by the King". This was the first model. Another model was made to a scale of one inch to a foot, and Hooke recorded seeing this with Sir Christopher Wren on 21st February 1674 and walking through it. This model is now in the crypt of St Paul's. More changes were made, as can be seen in the display in the crypt of St Paul's.

On 5th June 1675 the entry in Hooke's diary reads, "At Sir Chr. Wrens. He was making up my principle about arches and altering his module by it."

This is the principle of the Catenary Curve, illustrated by the loop of a suspended watch chain or necklace made of a series of links, and expressed by him in Latin in the entry for 26th September 1675, "Riddle of the arch, *pendet continuum flexile, sic stabit Rigidum*". This enabled the dome of St Paul's to be made thinner and lighter than any made previously.

The entry for 4th September 1679 reads, "Walked with Sir Chr. Wren in the Park, told him of double vaulting Paul's with cramps between". The finished dome consisted of an outer and an inner dome, with a conical structure between the two.

Meetings were held with the master craftsmen who were to be responsible for the work. On 2nd June 1673 Wren and Hooke discussed the model with Marshall and Streeter. On 25th October 1673 Hooke dined at Wren's with Grove, a master plasterer, and Storey, a master mason, when the first directions for the model were discussed. Wren and Hooke went to St Paul's on 11th April 1674, where they met Oliver, a glass painter and master mason, and Marshall, another master mason, and signed Marshall's report for £1000 for the committee.

On 5th June 1674, the diary records that Wren promised Fitch to be a brick layer at St Paul's, and on 11th May 1676, Hooke read his contract to him.

There were discussions about the materials to be used. On 7th December 1674, "Mr Bar here about marble for St Paul's". 6th February 1675, "With Sir Chr. Wren his new way of raising ballast". 23rd August 1675 "With Andrews to Sir Chr. Wren about sand and rubble for St Paul's." On 18th November 1676, "Dined with Sir Chr. Wren. After dinner a stone merchant discoursed about Portland."

During the years covered by the first diary some of the workmen died, and on 6th April 1687, Hooke noted that Joshua Marshall, mason died. The footnote adds that there are four items in the St Paul's accounts for September 1678, where there are four bills from his executors for an unpaid balance of £1191 12s 1½d. The footnote also stated that his death was a disaster as his successor's work was not satisfactory.

Hooke was critical of bad workmanship, and turned off

Cash's bricklayer on 8th June 1678, and earlier, on 20th August that year he wrote, "At Garlick Hill, Knowles a botcher."

Hooke was responsible for paying the workmen, and the entry for 31st August 1678, reads, "At Lord Mayor's, he signed warrants, I paid workmen, and gave acquitance in the chamber for the receipt of £1500, but had only his ticket of promise to pay it. I went to Scotland Yard and delivered Sir Christopher Wren's tickets to the workmen."

He appears to have had authority to make important decisions, as on 22nd June 1674 he ordered the pulling down of Bartholomews Tower. On 19th September 1674, he agreed with Marshall for St Stephen's, Coleman Street for £770 for east and south sides. He prepared ground plans for churches and on 12th April 1675, "Left at Sir Chr. Wrens the 14 ground Plats Harry had drawn."

It is difficult to determine what individual authority he had for on 9th March 1676, "went to Sir Chr. Wren's. Agreed with Cartwright for Bow Tower for £2550", and on 11th April 1676, "Agreed with Marshall about St Brides church tower." Later, on 2nd October 1677 he went to St Brides about measuring the steeple. "Observed that the Column, Cole Abbey and St Brides Steeple exactly in line."

It is clear from the many entries that the two men worked harmoniously together, and rarely a difference of opinion between them. Hooke frequently dined with Wren, and tipped his footman 5s and his coachman the same amount. He also had a hobbyhorse made for Wren's son.

Wren and Hooke went by water to the Playhouse on 20th June 1674 to see *The Tempest*, at a cost of 3s. They also discussed matters of scientific interest, and on 5th September 1674 Hooke had permission from Wren to publish Wren's paper on the straight motion of comets. Wren told Hooke of his way of flying by kites on 11th February 1675, and they had a wager about the rebounding of balls. Hooke discussed this theory of springs with Wren as early as 21st September 1675, and later awaited Wren's approval for the theory. Hooke told him that his theory was from the ether.

Two weeks later, Hooke demonstrated this to the King. Wren joined Hooke in the forming of a new club to discuss philosophical matters, the first meeting of which was held on 1st January 1676. A week later Wren and Hooke were discussing logarithms, arithmetical instruments, musical instruments, bells and sympathetic motions, and Hooke spoke of his theory of sound. This was one of many attempts to form a club for those more interested in purely scientific matters.

On 22nd June 1675, Hooke wrote that he had directions for the building of the observatory in Greenwich Park. This was designed by Wren, Hooke was the site and cost surveyor responsible for its building. The King had given £500 for this, but Hooke exceeded this amount by £26. He used stones, bricks and timber from one of the towers of the Tower of London, which was being pulled down. He also gained more cash from the sale of barrels of gunpowder at the Tower, which were judged to be unusable.

The Royal Society and Sir Jonas More helped to provide the instruments, items for which the King had apparently not budgeted. As he had just bought an expensive ring for one of his mistresses he may have had cash problems. The two "2 seconds clocks" were designed by Hooke, and made by Tompion. The mural quadrant was made by Thompson. In the structure of the Observatory was a deep well which may have been included for making observations of stars as they passed the meridian, the same idea that Hooke had used at Gresham College, where he had installed a zenith telescope by which he could observe stars in daylight, the first person to do so.

Hooke's plan for the Monument was the one chosen by the City. The records in the Guildhall Library contain both Wren and Hooke's designs. Wren's had tongues of flame at intervals up the sides of the column Hooke's was a a plain Doric column, 202 feet high, the tallest in the world, and

202 feet from the bakehouse in Pudding Lane, where the Fire first started.

There is an interesting footnote in Burton's diary relating to Sir Christopher Wren's uncle, Dr Matthew Wren, Bishop of Ely. He had been particularly active against dissenters in the early 1640s and when the Civil War started, was imprisoned in the tower.

The footnote on page 381 reads, "Mr (afterwards Sir C.) Wren when dining with Mr Claypole (Oliver's son-on-law) was surprised by the Protector Oliver's coming into the room (without the least notice being taken) sitting down and eating with them; during the repast turning to Mr Wren he said, 'You have a relation who has been long in the Tower, he may have his liberty if he chooses it'. 'Will your Highness give me leave to acquaint him with what you say?' 'Yes', was the reply. Mr Wren went with great joy to the old Bishop of Ely; but his answer was, 'This is not the first intimation of the same kind, but I scorn to receive my liberty from a tyrant and usurper.' And he remained in prison until the Restoration set him free."

One of Hooke's relatives of the same name was one of Oliver's chaplains and walked in his funeral procession. The times were certainly those of 'The World Turned Upside Down', as described by Christopher Hill in his book of that title.

CHARLES II

5th May 1673	"At Scaramouche's [puppet show] at York House, present, the King, the Duke of York, Lord Ormond etc."
23rd May 1673	"To Whitehall with Chancellor at Sir R. Morays walked the King seeing me cald me told me he was glad to see me recovered asked for measurement of degree by water."
8th Feb. 1674	"Sir J. More told me of Flamstead at the King and Col."
11th Feb. 1674	"Titus about Picart. Dr Pell told me of meeting at Col Titus about commission of longitude."
25th Feb. 1674	"To meet Dr Pell at Col Titus on Wednesday at 2 p.m. Propounded my business of longitude."
6th Mar. 1674	"Patent for Flamstead and observatory in Greenwich Park."
7th Apr. 1675	"With the King and shewed him my new spring watch. Sir J. More and Tompion there. The King most graciously pleased with it and commended it far beyond Zulichems, he promised me a patent and commanded me to prosecute the degree."
10th Apr. 1675	"At Sir J. More who told me the King's message and that unless we made haste with the watch he would grant the patent."
17th May 1675	"With Sir J. More to the King who received the watch kindly, it was locked up in his closet."
18th May 1675	"Met the King in the Park, he showed watch, affirmed it very good."

19th May 1675	"With the King, gave him paper against the Frenchman. Received the watch back, proposed longitude by time."
31st May 1675	"I showed the King the double pendulum clock which he liked well. He discoursed of weather barometer."
6th June 1675	"Showed Sir J. More my double Pendule clock, he admired it."
25th June 1675	"At Sir J. Mores. He told me Zulichems watch wanted minutes and seconds."
3rd Sept. 1675	"Perfected Philosophical Scales to show the King. Equall spaces with equall weights, and equall spaces and equall weights."
15th Sept. 1675	"With Sir Ch. Wren to the King he spoke of spying me in the Drawing Room. Cald me to him told me my watch did very well that he had tryd it the day before and found it true to his pendulum clock to a minute."
5th Oct. 1675	"At Whitehall saw the King in the Gallery he spoke to me in the Park that weather had altered watch. Bid me show him experiment. I prepared Experiment of Springs for the King tomorrow."
6th Oct. 1675	"By water with Harry to Whitehall. Walked in the Park with Sir Ch. Wren. The King called me to him, bid me show him experiment Followed him through tennis court and into closet. Showed him Experiment of Springs. He was very well pleased. Desired a chair to weigh in. Recommended me to the business of shipping. Discoursed of Basin in the Garden. I told him of movable keel and keel to let fall.
	The King weighed himself before and after playing tennis, and Evelyn noted that he lost 5 pounds in weight."

27th Feb. 1677	"... that the King had said I was a very able and honest man."
22nd Dec. 1677	"A letter from Sir R. Southwell to call me to the King with pepper mites."
24th Dec. 1677	"To Sir R. Southwell with him to the King's bedchamber. Shewd the King moss seeds and pepper mites. He was very well pleased, talked of Tangier, told me of Sir J. Moreland's weather glass, and bid me make one for him."

SIR JOHN CUTLER

In his preface to *Micrographia*, Hooke praised the particular generosity and munificence of Sir John for founding a lecture for the promotion of the mechanical arts. The lavish praise given to Sir John for, "his diligence for the Corporation of the Poor, his honourable subscription for the rebuilding of St Paul's and his cheerful disbursements for the replanting of Ireland, show by what means he endeavours to establish his memory? By their works shall ye know them". Sir John's behaviour towards Hooke and others earned him a very different epitaph.

When Sir John proposed the founding of these lectures, the salary was to be £50 per annum. In view of this, the Royal Society reduced the salary that they had proposed giving to Hooke as Curator to the Society, from £80 to £30 per annum. In spite of the several wealthy Fellows belonging to the Society, the weekly subscription of 1 shilling a week did not meet the expenses, and a thorough survey made by Michael Hunter, *The Morphology of The Royal Society*, showed that many were lax in paying even this amount.

Sir John paid half a year's salary in advance, but this was the only payment received by Hooke during the rest of Sir John's life. Hooke made many attempts during the next twenty five years to obtain his back pay, but a settlement was not made until 1696, six years after Sir John's death. By this time, Sir John's meanness had become well known, and was the subject of a poem by William Wycherley, entitled, *In Praise of Avarice*. As Sir John was able to bequeath an income of £10,000 a year to his heir Lord Radnor, there was ample justification for such an epitaph.

The entries in Hooke's diary show his patience and persistence in pursuing his claims, and the ingenuity of Sir John in thinking of delaying payment. In spite of non payment of salary, Hooke continued to give the lectures. When he heard of Sir John's death on 16th January 1690, Hooke commenced an action against Sir John's estate, which was finally settled in his favour on 18th July 1693, Hooke's birthday.

Hooke's Dealings with Sir John Cutler, how to avoid payment:

3rd Nov. 1672	"Went with Mr Lamot to see Sir John Cutler, there till night."
4th Nov. 1672	"Drank wine at Sir John Cutler's, agreed not."
3rd June 1673	"At Garraways – Sir John Cutler promised me payment for three years."
17th July 1673	"Spoke with Sir J. Cutler, he promised payment."
18th Sept. 1673	"With Sir J. Cutler he promised me money by postman."
30th Oct. 1673	"Met Sir J. Cutler, he discovered himself a cheating rogue."
17th Nov. 1673	"Visited and dined with Sir J. Cutler at Knightsbridge."
24th Nov. 1673	"Mr Axe promised to speak to Sir J. Cutler for money."
4th Dec. 1673	"Read noe lecture for Sir J. Cutler."
10th Dec. 1673	"At home all morning writing of lecture for Sir J. Cutler. D.H. Mr Axe here noe money from Cutler as promised, he pretended all must be for new set up at grocers."
11th Dec. 1673	"Read Sir J. Cutler's lecture."
20th Dec. 1673	"To Sir John Cutler to Knightsbridge. Spoke with him."
29th Dec. 1673	"With Mr Axe spoke again about Sir J. Cutler's money."
5th Jan. 1674	"Spoke with Sir J. Cutler at Tooly's coffee house with Mr Axe. Cutler promised money but I know not when."

15th Jan. 1674	"With Sir J. Cutler. He promised to give his final determination before the 1st of March and talked of will, and before Mr Bolter told me that he owed me for between 3 and 4 years."
26th Jan. 1674	"Good hopes of Sir J. Cutler."
22nd Mar. 1674	"Sir J. Cutler promised money."
13th Aug. 1674	"Demanded of Sir J. Cutler my money. He fooled."
13th Sept. 1674	"With Sir J. Cutler."
14th Dec. 1674	"Sir J. Cutler promised me money before Christmas, upon his word."
17th Dec. 1674	"At Garways. Sir J. Cutler played the fool with me."
4th Jan. 1675	"A letter from Sir J. Cutler, villain."
12th Apr. 1675	"At Dr Whistlers with Sir J. Cutler at the Crown, where he huffed, told me he cared not for my anger."
21st Apr. 1675	"At Sir J. Cutler. Dind. Oliver a villain. Cutler fals."
10th Dec. 1675	"Appointed Johnson on Tuesday morning about Rider and Cutler."
16th Dec. 1675	"Met Sir J. Cutler who invited me to dinner on Sunday."

On 31st December 1676, Hooke noted that Cutler owed him for six years and six months, £325 out of a total of £1300 owed from other sources.

30th Nov. 1678	"At Jonathon's with Sir J. Cutler, Sir Philip Mathews, Dr Whistler, Mr Baker, were he before them all acknowledged himself some hundred pounds in my debt, promised to pay me every farthing soe soon as ever I would call on him with Dr Whistler."
31st Dec. 1679	"At Sir W. Petty delivered him a book of springs. Discourse about affairs of Sir J.

Cutler, he advised to consult Councell" [of the Royal Society].

The full title of William Wycherly's poem based on his knowledge of Sir John Cutler was, *In Praise of Avarice, To a Miserable Wretch who pretended to hate Vanity, more than he loved Money.*

Dr Johnson's definition of a patron was, "One who countenances, supports or protects. Commonly a wretch who supports with insolence and is paid with flattery", and was written with experience of both. The first part of this definition could apply to Dr Wilkins and Robert Boyle who encouraged Hooke in the early part of his career, the second part aptly sums up Sir John Cutler.

Hooke expressed the favourable outcome of his action against Sir John Cutler's estate in Latin: *Deo omnipotentea mea Aetatis Gracias.*

CONCLUSION

Nullus in Verba

The wide range of Hooke's interests, investigations and achievements justifies the title of "Leonardo of London". Although he was not the first to attempt some systematic investigations with the microscope, his detailed examination of a wide variety of objects from the plant, animal and mineral worlds revealed things about their structure and physical changes for which he envisaged uses, and in some cases applied to make new instruments. His discovery of the coloured rings in mica aroused Newton's interest and caused him to make a closer study of them, leading to a further understanding of the nature of light. The instruments which he made and illustrated in the book for the closer examination of Nature which, as he stated, add to our means of perception to see and understand the greater detail, and admire the handiwork of God, and His Omnipotence, showed his skill and ingenuity as a craftsman. His further speculations about the things he revealed also showed remarkable insight as to the possibilities of more developments.

His skill as a draughtsman is also shown in these illustrations, and is comparable to the anatomical drawings of Leonardo in their clarity and detail. He also complimented the engravers who made the plates for the illustrations.

He showed his ability as an architect in the buildings that he designed and of which he supervised the building, and ensuring the soundness of their construction by the new spirit level which he designed, and the care that he took over the materials used. The principle of the arch and the idea about the structure of the dome of St Paul's gave Wren ideas which helped towards the construction of his masterpiece.

His ideas about the improvement of watches and clocks, together with the skill of the master craftsman Thomas Tompion, led to the production of the finest and most accurate clocks and watches at that time, all of which are still highly prized. The same collaboration resulted in the improving of quadrants, sextants, barometers and thermometers. He also designed and made telescopes, helioscopes and selenoscopes. These no longer exist, but the observations he made with them do.

His investigations into fossils and rocks could have added to the science of geology had the collection for which he was responsible been kept and not dispersed or neglected at his death, and possibly maintained an earlier interest in that subject, and its further development.

His systematic method of weather observations and the instruments he designed for that purpose were notable developments, and led to further interest in that science.

His knowledge of anatomy and skill in dissection were shown in the experiments to discover the functions of blood, and how the nitrous substance necessary for life got into it. His dissection of a variety of animals, a flounder and a porpoise and the examination of the skeletons of lizards, fish and crocodiles made him think about comparative anatomy.

Hooke's portrait was painted by an artist called Bonust on 16th October 1674, and during his life hung behind the President's chair at the Royal Society, together with that of Robert Boyle, his first patron, and lifelong friend and correspondent. The picture no longer exists, unless in some private collection. Hooke's first diary came to light in a sale in 1891 of the household goods of Moor Hall, near Harlow. At Hooke's death his library was sold, of which there is a record. The instruments which he made may also have been sold, but the only record of them are the drawings which were preserved by Richard Waller and William Derham. Some have been acquired by the British Museum. A picture in a stained glass window did exit until recently, in the west window of St Helen's, Bishopsgate. There were the pictures of six of the worthies buried in the church, among them Sir Thomas Gresham, Archbishop Bancroft, Sir John Crosby, Sir Julius Caesar and Robert Hooke. Hooke was buried about the middle of the south aisle, but in 1892 the church authorities sought permission to relay the stone floor of the church, and the City sanitary authorities complained of the smell. All the remains of those buried in the church were exhumed, six were reburied in the church, but Hooke's remains were taken with the others, and now rest in boxes in the vaults of the City of London Cemetery at Wanstead. Photographs of Hooke's portrait in glass exist, and show him in a brightly coloured suit, such as he had made for him by Nell Young, and looking quite sprightly, as he was until the last few years of his life.

His other memorials in the City are the Monument, and if holes could be cut in the platforms which obscure the view at present, visitors could look up and see the stars in daylight, as Hooke intended. The other is the dome of St Paul's.

Visitors looking up at the dome from inside can also see Wren's memorial, "If you want to see my memorial, look round you." From outside, visitors see the dome built according to Hooke's principle of the caternary curve, as mentioned in the diary.

Hooke's memorials in daily use are the constant velocity joint, the development of the universal joint, the kitchen and bathroom scales, operating on Hooke's Law of springs, the iris diaphragm in cameras, the wheel barometer, the hygroscope used widely in museums, and for those who own a Tompion clock, many of Hooke's improvements in clockmaking are incorporated in it. One of the clocks which he designed and Tompion made for the Royal Observatory, those with a 14 foot pendulum, is in the British Museum, the other is at Holkham Hall in Norfolk. A wheel barometer which he designed and Tompion made is at Hampton Court Palace.

Together with Dr Mayow, and Dr Lower, fellow Old Westminsterians, he came near to discovering oxygen, but at a time when there was no means of isolating and testing it.

All these add up to a considerable achievement in furthering scientific developments. He lived up to the motto of the Royal Society, *Nullus in Verba*, not by word alone, and did not give a farthing for a cartload of hypotheses. He served the cause of science for more than forty years, as Curator to the Royal Society, Cutlerian Lecturer in the Mechanical Arts, and Gresham Professor of Geometry.

His monuments are a plaque thirty feet high up on the building in The High at Oxford, where he first worked as Boyle's assistant, a damaged stained glass window in St Helen's, Bishopsgate, and the Robert Hooke Department of Meteorology at Oxford.

The proper recognition of The Monument in the City, as his work, where most of his time was spent, and where his contributions to rebuilding the City were done, would be a fitting memorial to a scientific genius.

Postscript

An attempt has been made to sort through the brief entries in Hooke's Diary, and present them in a series of articles about some of his many achievements, to bring these out of obscurity, and to give him the recognition he deserves as an inventor of several devices that have since been modified for present day general use, and an acknowledgment of his contribution to the Royal Society, and to several sciences, in the early days of their development.

APPENDIX

Music and Sound

The entry in his diary for 2nd February 1673 reads, "a fair but close day, morn very red. Th. 1¼, Hg. 125, Wind N.W. hard frost. Rose about 5 and slept till 9 pretty well. Sir Jonas More and Cox here; compleated arithmeticall engine in the contrivance for a product of 20 places. In the afternoon I found myself much better spitting much and also voyding by the nose but the following night slept very little and disturbed, but I sung much which made me not hear the noyse in my head."

Hooke had probably enjoyed singing from his boyhood days, in his father's church at Freshwater, subsequently at Westminster, and at Christ Church, Oxford. He attended a music house on several occasions, and noted that there was a new singer at Garraways, so musical entertainment was provided. On 27th September 1673 he was at a "musick house" with Mr Boaz, but did not state where it was. There is a record in *Brewer's Dictionary of Phrase and Fable* of the musical small–coal man, Thomas Britton of Clerkenwell, who established a musical club above his shop, which met every Thursday night. His dates are given as c. 1654–1714, the first date being too early for him to have established his business, but such entertainment seems to have been popular.

He had found that middle C had 272 beats a second, and on 1st September 1672 he noted that he had invented an easy way for "a musical cylinder with pewter tipes pinched between cylindrick rings".

Pepys recorded in his diary how Hooke told him that he could determine the number of beats of a fly's wing by the sound. This was on 8th August 1666, when Pepys discoursed with Hooke about the nature of sounds. "That, I suppose, is a little too much refined: but his discourse in general of sound was mighty fine."

He had a long discussion with Wren on 15th January 1676 on sympathetic motion, how the bow makes the fiddle string sound, how scraping of metal, and the teeth of a comb made sound. Compared light and sound and shewed how light produced colours in the same way by confounding the pulses."

He found that sound travelled along a wire, and that it could be made to go round corners. He was interested in echoes, as was shown by the design of Montague House, and the conveying of sound from one side of the room to the other.

He invented an ocousticon, a form of ear trumpet for assisting the deaf. This was in March and April, 1668.

On 8th July 1680 he formed the experiment of glass vibrating 6.4.8. places. This was done by putting flour on a glass plate, and bowing on the edge of the glass. The footnote states that Hooke had observed that the motion of the glass was vibrative perpendicular to the surface of the glass, and that the circular figure of the flour changed into an oval one way, and the reciprocation of it changed it into an oval the other way. This phenomenon was rediscovered by Chladni in the eighteenth century, and given his name. Its importance is that it influenced Faraday in thinking about lines of force in magnetic and electrical experiments.

Pepys wrote of going to the "great musique house", the King's Head, on 27th September 1665, but Hooke's entries were not as informative.

Shipping

As mentioned in Evelyn's diary on the occasion when he met Dr Wilkins, Dr Petty and Mr Hooke at Durdans during their stay to avoid the Plague, they were engaged on problems related to improving the rigging of ships and other means of improving the stability of boats. Dr Petty's method was to have a double bottomed boat, with two keels alongside each other. Such a boat was made, and christened *The Experiment*, and proved quite successful.

Hooke's method of giving greater stability was to have a

second false keel, which could be lowered in stormy weather, and raised in calmer. Having mentioned this idea to the King on 6th October, he continued with this as noted in the diary entry for 12th October. "Contrived about the movable keel of a ship. The great convenience of it for making a man-of-war stiff by the weight of the said keel and by keeping the ballast to the right side where it ought to be. How it will strengthen a ship to have a deck immediately over the Ballast. Resolved to prepare module of ship with movable keel."

On the 16th he noted, "Sir J. Morelands huff about the center of Gravity of ships."

On 18th December he, "discoursed with Cap. Sheeres and 2 other ingenious men about ships, about Bond's Theory of the motion of the magnetical vertue within the body of the earth, or the Universal Character of Navigations and navigators."

Three years later he told Sir J. Hoskins of my double keeld ship, etc. His entry for 2nd February was that he was at Jonathon's where he, "discussed with Sir John Hoskins his little Ship, Sailes, mast steering, sinking, pumping, till 10".

On 4th February, he noted, "made ship, on 3rd March, Ship Launched, and on the 9th reported this to Sir John Hoskins".

Later in the year, on 29th September he went with Sir John Hoskins to see a ship module at Christ Church. Whether this was Hooke's model or another one cannot be sure of from Hooke's brief notes.

Another statement made while at Garaways on 9th July was that "Straight ships always spreading upward square to the Rudder. Great use may be made of a windmill at sea for a chaine pump for Raising water, For Raising the anker in a shalop or the ship, for grinding, for Raising sayles or Goods, for winding up in a creek against the wind by an anker and cable".

How sound these ideas were only seafarers would know, but entries in the Everyman edition of Cook's voyages, p. 75 and 100 show that his ship was fitted with a false keel.

Details provided by the Science Museum from the History of Naval Architecture by Fincham give details of two ships which had false keels, p. 248.

An article in *The Independent* in 1986 mentioned a ship, *HMEICS Nemesis*, launched in 1840, and made by Laird's of Birkenhead which had two lifting keels, and which proved to be very seaworthy in rough weather. A letter from Derek Wakern following this article gave more details, so the idea was put into practice, and proved Hooke's theory.

Hooke also invented a waywiser for use at sea, but there are no details.

A Perfect Wheel Work

On 31st October 1667, Hooke produced two instruments of his own contrivance, one called a perfect-wheelwork, so made as equally to communicate the strength of the first wheel to the last, so as before one tooth had done taking, it was passed a good way into another. This method also made worm gears possible, which gave a high velocity ratio.

New Kinds of Levels

On 28th November 1666, Hooke produced a new kind of level by including a bubble of air in a glass pipe, having its sides exactly blown, and filled with water, and sealed up at both ends.

On 5th December 1666 Sir Robert Moray mentioned a new kind of level contrived by Dr Christopher Wren, which Mr Hooke was ordered to get made as soon as possible.

On 12th December, Dr Wren's level being called for, it was produced ready made, and ordered to be described. From the description given on 14th March 1667, Wren's was not a sealed glass level, as was Hooke's described on 5th December 1666. He did produce a level almost the same as that of the French, of which an account had been given in the *Journal des Scavans*.

As both men were then engaged on the task of rebuilding the City of London they would have been interested to ensure that the levels of brick and stonework would be accurately laid.

An Alphabet of Symbols

On 25th February 1669 Dr Holder presented his written discourse concerning the Elements of Speech, an *Essay of Inquiry into the Natural Production of Letters*, together with an appendix to instruct persons deaf and dumb. The alphabet which Hooke devised, and which is illustrated might have been in response to this discourse, but it could have been the stimulus to the inventor of the Braille Alphabet, where the spots are raised in their distinctive patterns so that the blind or partially sighted can feel and discern which letter is which.

The Universal Joint

On 18th September 1676, Hooke bought brass plate at 1s 6d a pound. On the following day he wrote about the joint, and on the 21st filed and drilled it, and completed filing the joint for an instrument on the 25th.

He "contrived motion of pendule without noyse by universal joynt on 29th June, 1677".

This device, first made to allow a pendulum to move noiselessly, is now an essential part in the transmission of all cars, lorries, and mechanically propelled vehicles. Many experiments and alterations had to be made before it was developed into the present constant velocity joint, one man responsible for this being M.J.A. Gregoire, French racing driver and engineer who died on 20th August 1992, and whose obituary appeared in *The Independent* on the 24th.

In Britain it is known as the Hardy Spicer joint, and is made in large quantities by G.K.N.

The largest universal joints are those on the feet of oil rigs, where they are as large as a London double decker bus, and allow the rig to move slightly according to wave and current movements.

A Device for Speedy Intelligence

On 17th February 1663, Hooke was asked to bring in his report on his apparatus for the management of speedy intelligence.

This report did not appear until the meeting of 29th February 1671. At this meeting he proposed a way for a very speedy conveyance of intelligence from place to place by the sight assisted with telescopes, to be employed on high places, by the correspondent using a secret character, proportioned in bigness according to the distance at which they are to be seen. That a system of sending messages rapidly from one place to another was eventually used is by the fact that semaphore signals were sent from Brussels to Britain to give the news of the victory at Waterloo, and commemorated at the high places with the name Telegraph Hill.

During the wars against the Dutch, Pepys had commented on the necessity for speedy intelligence.

Hooke and Education

There were two comments about education in the diary. One was the suggestion of teaching grammar by tables, which might appeal to some. The other was teaching languages by exchange, a very sensible idea for modern languages. In some schools at the time only Latin could be spoken during school hours. Hooke learnt Dutch from Blackburn in order to be able to correspond with Leeuwenhoek, the Dutch microscopist, and this correspondence is related in full in the biography of Leeuwenhoek published by the Dover Company.

Hooke read John Locke's *Defence of Human Reason*, and bought his book on education. This went through eighteen editions in England, and sold widely in France, Germany, Italy and Switzerland. One or two comments show his approach to education. "Curiosity should be carefully cherished in children, as other appetites are suppressed. Labour for labour's sake is against nature. There is but one fault for which children should be beaten or punished and that is obstinacy."

Hooke was fortunate for his day, as during his time at home in the Isle of Wight he was allowed a great deal of liberty, and excused the rigid study usually served to children. When at Westminster he again was allowed to study by himself, and was not subjected to Dr Busby's severe discipline, rather like the ideal advocated by Francois Rabelais at the Abbey of Theleme, with its motto, *Do what thou wilt*. John Aubrey was also of the opinion that stimulating learning by beating stifled the imagination.

John Locke's book on education is available in an English edition, Aubrey's views have to be gleaned from the detailed introduction to his *Brief Lives*, edited by Oliver Lawson Dick.

Lamps

On 19th October 1675, Hooke had observed the detail of a flame, and its having inner and outer cones. This led to him making further studies during that year and into 1676, and devising a lamp which would give a constant supply of oil to the wick, and also having an adjustable "poyse" for the lamp. These results were published in his paper, *Lampas*.

Flying

This was the topic of conversation during the years of the second diary from 1672–1680. On 18th December 1675, he considered making flight possible by means of an airscrew. He had earlier, on 7th November 1673, thought of using powder, one supposes as a propellant. One of Sir Christopher Wren's suggestions was that flying was just like walking up a flight of stairs at 45 degrees. There are more than twenty references to this subject:

7th Nov. 1673	"To Spanish coffee house with Lingar and Blackburne. Discoursed divers ways of flying by powder, & c."
4th Oct. 1674	"Tompion here all day. Discoursed of the way of bending springs by gunpowder for flying."
7th Oct. 1674	"First tryd experiment about artificial strength by water, air fire, by which flying is easy and carrying any weight."
8th Oct. 1674	"At Councell at Arundel told Sir Robert Southwell that I could fly, but not how."
11th Feb. 1675	"Dr Croon at Royal Society read of the muscles of birds for flying. I discoursed much of it. Declared that I had a way of making an artificial muscle and to command the strength of 20 men. Told my way of flying by vanes tryd at Wadham. Told Dr Wren's way by kites, of the unsuccessfulness of Powder for this effect, and what tryalls and contrivances I had made."
9th Aug. 1675	"Invented flying chariot."
25th Dec. 1675	"Read Lord Chester (Dr Wilkins) about flying & c. Combined flying engine."
28th Dec. 1675	"Directed Shortgrave in Blackfryers about flying Engine."
30th Dec. 1675	"With Mr Wild at Joe's. Described to him my contrivance of flying."
9th Jan. 1676	"At Garaways. Discoursed with Harry about my undertaking to fly."

11th Jan. 1676	"Contrived flying by pulleys without wheels."
28th Mar. 1676	"To Mr Boyle. Roberts there. Propounded to him about flying."
30th Mar. 1676	"D.H. Noe meeting of the Royal Society. Sir John Hoskins. Discoursed about my flying and rowing engines, the use of it in chariots, water drawing, rowing & c., about dancing shoes."
14th July 1676	"At work in turret about Helioscope and flying pump."
31st Dec. 1678	"Read Mr Barnard's Letter at Mr Haak's of one that had a way to fly to Paris."
1st May 1679	"Licensed flying discourse."
31st May 1679	"Told Sir Ch. Wren of flying module."

In her biography of Sir Jonas Moore, Frances Willmoth quotes the story related by Aubrey about Sir Jonas, in his *Brief Lives* of, "a Jesuite (I think 'twas Grenbergus, of the Roman college) found out a way of Flying, and that he made the youth performe it. Mr Gascoigne taught an Irish boy that way, and he flew over a river in Lancashire. But when he was in the ayre, the people gave a shout, wherat the boy being frightened, he fell downe on the other side of the river, and broke his legges, and when he came to himselfe, he sayd that he thought the people had seen some strange apparition, which fancy amazed him. This was ano 1635, and he (Sir Jonas), spake it in the Royall Societie, upon the account of the flying at Paris, two years since."

This was registered in the Society's Journal for 8th May 1679.

Sir George Cayley, one of the pioneers of flying, had his coachman pilot a glider across a river in Yorkshire in about the year 1830, and this made the journey successfully. Sir George's researches into the basic principles of flying were acknowledged by the Wright brothers, who first used a sufficiently light power unit to lift their plane off the ground.

Hooke had bought a book of Leonardo da Vinci's works on 29th October 1673 which cost him 15s, so may have been familiar with Leonardo's ideas on flying. As with the early work on blood transfusion, this took two more centuries before success was achieved.

Two months before Hooke bought the book on Leonardo, Sir Jonas Moore had visited Hooke, and promised £100 for books for the library of the Royal Society, and the gift of a brass sphere.

Some Random Experiments Not Pursued

22nd June 1671. The experiment for showing the internal motion of liquids was made, by putting some small pieces of charcoal into spirit of wine in an open glass, which being viewed through a large microscope appeared to have a very vehement motion every way, though to the naked eye there appeared none. Hooke said that there was no such motion in common waters or vinegar and that he was of the opinion that all spirituous liquors would exhibit such a motion.

In 1827 a Scottish botanist, Robert Brown, demonstrated the movement of molecules in liquids with pollen grains in water. Hooke did discourse with Hill about the particles of body, figures of ice, frost and snow on 25th March 1676, but that seems to be the last mention of it.

On 20th June 1667, Hooke tried the experiment, which he had been instructed to do at the meeting on 6th June, of taking away the sharpness of vinegar, and reducing it to a real sweetness, by putting in a small amount of red lead. The sharpness of the vinegar was much reduced. This made "sugar of lead", sweet, but highly poisonous. Hooke repeated the experiment of trying to reduce the sharpness of vinegar with lead filings, eggshells, brass, steel dust, and oyster shells, all of which deprived the vinegar of its acidity. Crab's claws and chalk were also used. These experiments were first suggested by a letter from Dr Pope to Hooke on ways of making groat ale, and reducing its sourness.

On 18th June 1668, Dr Wilkins moved that Mr Hooke might be ordered to try whether he could by the means of the moss seed shown by him make moss grow on a dead man's skull. There is no record of the success of this.

A Few Suggested Cures of that Time

Gravel	"Mr Boyle's cozen, a Quaker, told me a most soverain medicine for the gravell, twas a powder made of Darbyshire Sparr which floats like crystals."
Stone	"19th Nov. 1672. Sir Theodore Devaux told me of Sir Th. Mayerns cure of stone in kidneys by blowing up bladder with bellows."
Sciatica	"At Sir Jonas More, sick, he was cured of a sciatica by fomenting the part for an hour with hot steames afterwards chafing in oyles with a rubbing hand and heated firepans, which gave him sudden ease."
Collick	"13th Jan. 1673. Mr Axe told me of my Lady Portman relieved the greatest of collick paines by a clyster of Venice treacle."
Leprosy	"31st Jan. 1673. Sir A. King told me of a kinswoman of his which has a certain cure for leprosy or Scale head. She had £100 per annum of St Bartholomews hospitall to which she promised to leave the receipt."
Jaundice	"13th Jan. 1674. Mrs Tillotson told me of an ounce of castile soap boyled in a pint of ale till ye ale was half

	consumed and dranke warm was a sure medecine for the yellow Jaundice which had been often tryed with certain effect."
Consumption	"3rd Feb. 1674. Mr Hedges told me the flesh of tortoises any ways eat is a certaine cure of the consumption."
Mad dogs	"24th Feb. 1674. Mr Hoskins told me of King's bit by mad dogs cured by a felon in the neck."
Piles	"22nd Apr. 1674. In St Martins for geese greese exceeding good for the piles. Being made into a supposition sold for 5s per lb. Another cure for this ailment was given by Lady Ford, who had a sanatine Drink communicated to her by Dr Hamey. Nell told him of another cure which was made of oyle of roses, powder of myhr, powdered frankinsense, honey, one pennyworth of each, mix and anoint. Another cure was the application of horse leeches."
Staunching blood	"5th Feb. 1674. Hooke tryd Mr Lister's new Liquor for staunching blood. Later, on 30th Dec. 1675, Hooke cut off the top of his thumb, but cured it in four days by Balsamus Peruvianum supplied by Shortgrave."
Morphea	"1674. A ready way to cure the morphea is by sweating in a stove and washing the place with oyle of tartar."
Chilblains	"26th Nov. 1674. Mr Wild told me that the blood of a black cat will cure chilblains."
Rheum in the eyes	"13th Aug. 1675 Dind at Mr Wild. To stop a rheum in the eyes use flowers of brimstone and milk."

Toothache	"5th Feb. 1676. Camphire from AF. (aqua fortis) good for toothache.
Giddiness	"22nd Sept. 1676. At Garraways. Knox told of medicine for giddiness, either to smoke coltsfoot in a pipe or put it in a broth."
Falling sickness	"17th Dec. 1676. Mrs Tillotson told me also a soveraine remedy for falling sickness was made out of the mosse of a man's scull."
Wormes and Thrush	"16th Aug. 1677. At the Crown, Sir Christopher Wren told me of killing wormes with burnt oyle, and curing his Lady of a thrush by hanging a bag of live boglice about her neck."
Bad memory	"At Garaways Mr Melancholy told me that a friend of his had recovered of a bad memory and several other distempers by carrying a small box full of very fine filings of the best refined silver, and now and then licking of it with his finger and swallowing it."
11th April 1681,	Letter from Cesare Morelli to Samuel Pepys,

"Honoured Sir,

I did receive your last letter, dated the 9th of this month, with much grief, having an account of your painful feaver: I pray God it will not vex your body too much; and if by chance it should vex you longer, there is a man here that can cure it by sympathetical powder, if you please to send me down the pearings of both your hands and foots, and three locks of hair of the top of your crown. I hope, with the grace of God, it will cure you."

This was an example of making the powder of sympathy, and at the meeting of the Royal Society on 23rd December 1663, Christopher Wren had related the experiment of trying to cure a maid servant's cut finger by the powder of sympathy method, but his opinions as to the effectiveness of such treatment was not stated.

On 21st July 1677, Hooke wrote in his diary, "Lord Tannet killed by Young's pills", but did not state the composition of these pills.

A few cures mentioned in Aubrey in his *Brief Lives*. He mentioned how the reading of Sir Thomas Browne's *Religio Medici* first opened his eyes, and that from that time his learning suddenly took a leap forward. Browne's opinions as expressed in his book *Vulgar Errors* was that there was too much adherence to antiquity and tradition, authors filching other people's ideas, and wishing that men were still not content to plume themselves with other people's feathers.

He also thought that testimony was of small validity if deduced by men out of their profession. Aubrey mentioned that the medical practice of Dr William Harvey, discoverer of the circulation of the blood fell off after the publishing of his book describing his experiments and theory. In contrast to this Aubrey mentions a Dr Napier, who was no medical doctor but practised physic, who, when a patient came to him for advice and treatment, went into his closet to pray and converse with the Angel Raphael, who told him whether his patient's ailment was curable or incurable, and that the popularity of this treatment was proved by the fact that Dr Napier's knees were horny with frequent prayer.

Aubrey related a case where a beloved daughter had received no benefit from physicians whom she had consulted. Her mother dreamed that a friend of hers who had recently died, told her in a dream that yew leaves crushed and boiled in water would make a medicine to cure her daughter. The mother made this brew and gave it to her daughter, who died. The housemaid was so upset by the death of the daughter that she said there must have been a mistake, and that something other than the brew had caused her death. To prove it, she took the brew and also died.

Aubrey related of a case where a woman in Italy wished to poison her husband, and boiled a toad and put it in his potage. This improved his health, and became another cure of those times. A cure for toothache given by Aubrey was to take a new nail, scratch the gum with it to make it bleed, and then drive the nail into an oak tree. "This did cure William Neal, Sir William Neal's son, who was almost mad with toothache."

Pepys wrote in his diary that he carried a hare's foot to prevent attacks of colic, and took pills made of turpentine to avoid stones in the urinary system. He had been cut for the stone in his bladder on 26th March 1658 by a method which probably also severed the seminal ducts, which made him sterile. He kept the stone in a box, and celebrated the success of the operation every year. Dr Wilkins, one of Hooke's patrons, is in some accounts stated to have died as the result of stones in his bladder, but Hooke in his diary gave the cause of death as suppression of urine, probably due to enlarged prostate gland, and that the autopsy revealed only a few small stones in the ureter.

On 13th April 1661 Pepys had seen the King touching people for scrofula, the King's Evil. Dr Guthrie in his *History of Medicine*, states that Charles touched on average 4,000 people a year, and a total of 92,107 during his reign.

The examples of the cures that Hooke tried, and the few that he noted and are given here, would present a one-sided view of medical knowledge and the advances made in that subject. Among the physicians and doctors mentioned in the diary were some who made notable discoveries. Dr Willis, for whom Hooke acted as an assistant in his studies in chemistry, and who taught Hooke dissection, made researches into the nervous system, and his book *Cerebri Anatome*, one of the first descriptions of the brain, was illustrated by Christopher Wren. The arrangement of the arteries at the base of the brain is known as "The circle of Willis". He also noted the phenomenon of hearing better against a background of noise, and is credited with the discovery of sugar in the urine.

Dr Francis Glisson wrote a detailed description of the liver, and investigated rickets in children.

Dr Thomas Sydenham is considered the greatest clinical physician of the seventeenth century, and is referred to as the English Hippocrates. He had no time for book learning, although he had studied at Oxford and Montpelier. He considered that the only place to learn about disease was at the patient's bedside. Dr Thomas Dover, one of his pupils, related how Dr Sydenham cured him of smallpox by forbidding a fire in his room, keeping the windows open, the blankets and sheets no higher than his waist, and drinking twelve bottles of small beer every twenty four hours.

Dr George Thompson, who is mentioned in the diary, made an attempt to discover the cause of the plague by post mortem examination. Dr Nathaniel Hodges stayed in London during that time and wrote a detailed account of his daily visits to his patients, and his precautions to avoid being infected.

Dr Richard Wiseman, the "Father of English Surgery", served with the Royal army during the Civil Wars. Afterwards he was appointed surgeon to King Charles II, wrote a book *Several Chirurgical Treatises*, and a detailed account of scrofula, *King's Evil*. Like the famous French surgeon of the previous century, Ambrose Paré, Wiseman wrote a clear account of gunshot wounds and their treatment.

There are two references in Hooke's diary to Sir Thomas Browne, whose book *Vulgar Errors* Hooke bought for 8s 6d from Martin on 6th August 1678, and his *Urnotophia* on 17th January 1679, from the same bookseller. Sir Thomas Browne was born in London in 1605, and educated at Oxford, Padua, Leyden and Montpelier. He practised medicine at Norwich, and his best known book is *Religio Medici*, which Aubrey read, and which aroused his interest in learning. He had a critical mind which would not accept the literal interpretation of the scriptures, for he admitted that for him the first book of Genesis had a great deal of obscurity. His book *Enquiries into Vulgar Errors* was a plea for using Baconian methods in research.

A few examples from this book show how fanciful were some ideas prevalent at that time.

In Chapter VII of the book he states, "There are surely few that believe or hope enough to experiment the Collyrium of Albertus; which promiseth a strange effect, and such as thieves would count inestimable, that is to make one see in the dark; yet thus much, according to his receit will the right eye of a hedgehog boiled in oyl, and preserved in a brazen vessel effect. As strange it is, and unto vicious inclinations were a nights lodging with Lais, what is delivered in Kiranides; that the left stone of a weasel, wrapt up in the skin of a she mule, is able to secure incontinency from conception. That the body of a man is magnetical, and being placed in a boat, the Vessel will never rest untill the head respecteth the North. If this be true the bodies of Christians do lye unnaturally in their graves. Lastly, that a beaver to escape the hunter, bites off its testicles or stones, as mentioned by Aristotle in his Ethics, and in Pliny and Juvenal."

There are two remarkable operations dating from these times. One made John Locke famous: Anthony Ashley Cooper was showing slight signs of having jaundice, and had a hard lump below the lower ribs on his right hand side. This was diagnosed as a growth on the liver. An incision was made below the ribs, and the growth cauterised. The wound was irrigated with weak wine and a wax candle inserted in the wound. This was removed the following day covered with cysts and pus. The wound was again irrigated and a larger candle inserted.

On the next day this was removed covered with pus. The wound was cleaned in this manner several times, and then a metal tube was inserted to drain away any more fluid. Ashley Cooper retained this tube in his side for the rest of his life, and it was known jokingly as "Shaftesbury's Tap", as he later became Lord Shaftesbury, a member of the

Cabal, and one of the most influential politicians of his day. Locke, who was present at the operation, became his personal physician. Hooke showed Lord and Lady Ashley round the repository of the Royal Society on 13th August 1672.

The other operation was one reported to the Royal Society, and recorded in Birch's *History of the Royal Society*. M. Borel of Paris wrote a letter in which he related that a man had been wounded in the chest in a duel and left for dead. He was treated by a surgeon, and the wound healed, but the man's health deteriorated, and he grew weaker. A M. Suif, a renowned surgeon living in the reign of Louis XIII, after hearing details of the man's injury and condition, said that he would cure him, if he could endure the pain being inflicted on a man being broken on the wheel sixteen times. The man agreed, and was taken into the surgeon's house. M. Suif made an opening in the chest wall large enough for his hand to be put in. He then pulled the lung out towards the opening and cut off the diseased part of the lung with scissors. The patient said that he could not endure further pain, but was persuaded by his friends to continue with the treatment. After the sixteenth day the lung was not touched any more, and the opening was covered with a metal plate. The patient was afterwards in good health, and lived for another ten years. He was able to live a normal life, but was a bit short-winded. This letter was read to the Royal Society on 15th February 1665.

While these two examples of surgery met with success, Samuel Pepys, on the advice of Robert Boyle, sought the opinion of Dr Daubeny Turberville, the most famous eye specialist in his day. This man had not seen an eye dissected until 1668, and treated Pepys with eyedrops and purges. Hooke had made an artificial eye which he showed to the Royal Society on 12th and 14th October 1663. This included the iris diaphragm, which he had devised by overlapping plates, and which could be opened or closed by a string at the side.

Buildings designed by Hooke. Still in existence:

Buntingford, Herts, Bishop Seth Ward's Almshouses, built 1689.
London, Hoxton, Haberdasher Aske's Hospital and School, 1690–3.
London, The Monument, Fish Street Hill, 1671–6.
Westminster School, Dr Busby's Library, damaged by bombing, 1942, and rebuilt.
London, Burlington House, Piccadilly, 1676–77 for Lady Burlington.
London, Greenwich Observatory, 1675.
Lowther, Cumberland, alterations to the church, for Sir J. Lowther, 1690.
Alcester, Warwicks. Ragley Hall; extensive alterations for 1st Earl of Conway.
Willen, Bucks, Willen Church for Dr Busby.
Ramsbury Manor, Wilts., for Sir William Jones, 1673
Shenfield Place, Essex, for Richard Vaughan.
Plymouth, Royal Dockyard, Hamoaze, The Officers Dwelling Houses, Great Storehouse and Ropehouse, 1690–1700. The Dictionary of British Architects has a question mark by this entry, and notes that a set of designs for these buildings, as engraved for the benefit of the Navy Commissioners is preserved among the Hooke papers in the British Museum, and that these closely resemble the Bethlehem Hospital in style.
Cambridge, Magdalene College, the Pepys' library, which was under construction in 1679, and for which there are references to Pepys in 1677.
There are many references for work for Lady Portland between December 1678–1680.
Similarly for Lord Bath in 1678.
Similarly for Lord Manchester in 1677.
Similarly for Lady Ranelaugh in 1677.
Similarly for Lord St Albans in 1678.
Dr Frances Willmoth, in her doctoral thesis on Sir Jonas Moore (1992), and subsequent book of the same title, states

on pp. 170–171, with respect to Greenwich Observatory that the architect for this was Robert Hooke, and that "It is not uncommon for architecture which was really Hooke's to be attributed to Wren."

As her supervisor for her doctoral thesis was Dr. J.A. Bennett, whose own doctoral thesis was on "Studies in the Life and Works of Sir Ch. Wren" dated 1972, Cambridge, there are sound reasons for acknowledging the design of Greenwich Observatory to Hooke.

A feature of the design of Greenwich Observatory was the adaption of the former well of the castle to make a zenith telescope, an idea which Hooke had used at Gresham College, and in his design of the Monument, and by which means he had been the first man to observe a star in daylight.

No longer in existence:

Escot House, Devonshire, designed for Sir Walter Yonge, altered by James Wyatt, destroyed by fire in 1808.
Londesborough House, Yorks, designed for the 1st Earl of Burlington. Demolished in 1811.
Bethlehem Hospital, Moorfields, 1675–6, demolished 1815–6.
The Writing School, Christ's Hospital, Newgate Street.
Merchant Taylors' Hall, hall screen, destroyed by bombing 1940–1.
Merchant Taylors' School, Suffolk Lane, London, 1674–5, demolished 1875.
Montagu House, Bloomsbury, for Ralph Montagu, cr. Duke of Montagu, 1675–9. Gutted by fire Jan. 1685–6, rebuilt 1687, apparently by the French architect Pouget, and from Celia Fiennes' description before and after the fire, using the original walls. Hooke's forecourt and gateway survived until the nineteenth century. It was the first home of the British Museum.
The Royal College of Physicians, Warwick Lane, 1672–8.

The domed octagonal theatre was demolished in 1866, and the remainder of the building destroyed by fire 1879.

Nos 6–7, St James's Square, London, for John Hervey and the 1st Earl of Ranelagh, 1676–7. Demolished.

The Stables, Somerset House, Strand, London, for Queen Catherine of Braganza, 1669–70. Demolished 1780.

A house in Spring Gardens, London, for Sir Robert Southwell, 1684–5.

5 houses in The Strand, London, for John Hervey, 1678.

A house in Privy Gardens, Whitehall for the 20th Earl of Oxford, 1676, demolished 1691 or 1698.

Canterbury Cathedral, panelling in the choir, 1676, removed 1826.

A new Bridewell, at Blackfriars, where the Fleet River enters the Thames.

The Fleet River Quayside.

A new Conduit at Holborn Bridge.

A house for Mr Gold in Highgate Village, possibly Bisham House.

SELECT BIBLIOGRAPHY

Birch, Thomas. *The History of the Royal Society*, Edition Culture et Civilisation, Bruxelles, 1968

Braybrooke, Lord. *The Diary of Samuel Pepys*, Odhams 1934

De Beer, E.S. (ed). *The Diary of John Evelyn*, Oxford, 1959

Dick, O.L. (ed). *John Aubrey, Brief Lives*, Secker & Warburg, 1949

Gunter, R.T. *Early Science at Oxford*, vols VI, VII, VIII, IX, X, Oxford, 1935

Guthrie, Douglas. *The History of Medicine*, Nelson, 1930

Hooke, Robert. *The Diary of Robert Hooke*, Taylor & Francis, 1935

Hooke, Robert. *Micrographia*, Edition Culture et Civilisation, Bruxelles, 1968

Hope Nicholson, Marjorie. *Pepys' Diary and the New Learning*, Virginia University Press, 1965

Hunter, Michael. *John Aubrey and the World of Learning*, Duckworth, 1974

Hunter, Michael. *The Morphology of the Royal Society, 1660–1700*, B.S.H.S., 1982

Hunter, Michael. *Science and Society in Restoration England*, Cambridge, 1981

Ogilby, John. *Brittania*, Duckham, 1939

Reddaway, A.F. *The Rebuilding of London after the Great Fire*, Jonathan Cape, 1940

Symonds, R.W. *Thomas Tompion*, Batsford, 1951

Waller, William (ed). *The Posthumous Works of Dr Robert Hooke*, 2nd edn, Frank Cass, 1971

Waller, William (ed). *The Philosophical Works of Dr Robert Hooke*, 2nd edn, Frank Cass, 1971

Westfall, Richard. *Never at Rest – A Biography of Isaac Newton*, Cambridge, 1980